ANIMAL SCIENCE, ISSUES AND PROFESSIONS

FELINES

BEHAVIOR, CLASSIFICATION AND DISEASES

Animal Science, Issues and Professions

Additional books in this series can be found on Nova's website under the Series tab.

Additional e-books in this series can be found on Nova's website under the e-book tab.

ANIMAL SCIENCE, ISSUES AND PROFESSIONS

FELINES

BEHAVIOR, CLASSIFICATION AND DISEASES

RODRIGO E. MORENO GARCÍA
AND
RUBEN G. RUIZ JIMÉNEZ
EDITORS

Nova Science Publishers, Inc.
New York

For permission to use material from this book please contact us:
Telephone 631-231-7269; Fax 631-231-8175
Web Site: http://www.novapublishers.com

Library of Congress Cataloging-in-Publication Data

ISBN: 978-1-62257-005-8

Published by Nova Science Publishers, Inc. † New York

CONTENTS

PREFACE

In this book, the authors gather and present current research in the study of the behavior, classification and diseases relating to the feline species. Topics discussed in this compilation include the use of Type 1 interferon for treating feline retrovirosis; new advances in the diagnosis, epizootiology and taxonomy of feline piroplasmids; feline hepatozoonosis and the resurrection of hepatozoon felis; the role of cats in toxoplasma gondii infection; and the therapeutic effect of melatonin in feline uveitis.

Chapter 1 - This chapter reviews the pathogenesis of the infections by feline immunodeficiency virus (FIV) and by feline leukemia virus (FeLV) as a way to understand the clinical signs. Even though there are commercial vaccines available, the reality is that these are two of the most prevalent infections in many areas of the world. As most retroviral infections, they are difficult to treat because the virus integrates into the feline DNA, behaving as a host gene. An alternative or complement to antiretroviral therapy is the use of type I interferons. Interferons are low molecular weight glycoproteins synthesized by several cellular types, mostly of the immune system (macrophages and T-helper cells). They are found naturally in the serum of animals infected by viruses and represent one of the components of the intrinsic immunity. Due to their antiviral activity, their potential to boost the natural immune response, and to the clinical results obtained with HIV, interferons have been proposed for treating feline retrovirosis. However, their efficacy in this respect is controversial. Though long-term studies are lacking, it seems that in the short-term the clinical condition of cats improves transiently with type I interferon (IFN) treatment, but this is not associated with a decrease in viral or proviral loads. Studies on the in vitro effect of IFN to clarify whether or not IFN affects the retroviral cycle suggest that this chemokine may decrease the number of infective viral particles, but does not act directly on viral expression but rather on morphogenesis and budding. Reports on the lack of specific sequences in the control regions of the proviral genome (LTRs) of FeLV and FIV agree with these in vitro results. Most studies have been carried out on infected cats with no tumors, and the possible use of interferons for treating the different tumors which may accompany feline retroviral infections is still far from confirmed. Thus, more studies are still needed to understand better the effects of interferon treatment in feline retrovirosis.

Chapter 2 - In vertebrates, piroplasmid protozoa infect blood cells. Both wild felines and domestic cats may be infected by *Babesia* and *Cytauxzoon* species. Feline piroplasmosis is assumed to be tickborne, but the definitive host has not been identified in many piroplasmid species. Babesiosis and cytauxzoonosis are considered as emerging diseases, affecting felids in a wide geographic range. Usually these infections are asymptomatic but in some cases disease is evident and animals may eventually die. This is much more likely whenever concurrent pathogens (such as *Mycoplasma* or viruses) attack felids. It must be emphasized that babesiosis is present in wild felines, included those qualified as endangered species. Piroplasmids may be a severe threat for the Iberian lynx (*Lynx pardina*) and the Serengeti lion (*Panthera leo*). Although piroplamosis may be treated with primaquine, imidocarb or some other drugs, chemotherapy results are not always satisfactory.

Before the advent of molecular diagnosis methods, piroplasmid species infecting felines were limited to *Babesia cati*, *B. felis*, *B. herpailuri*, *B. pantherae* and *Cytauxzoon felis*. However, some contributions published in the last years have shown that there is a broader spectrum, which includes *Theileria annae* (or *Babesia microti*-like), *B. leo*, *B. canis canis*, *B.canis presentii*, *C. manul*, *Babesia* sp, *B. canis vogeli* and *B. lengau*. The classification of feline piroplasmids is in continuous evolution due to the application of molecular techniques such as DNA sequencing. In agreement with the opinion expressed by some authors, and based on current biological and epizootiological data, it is evident that the species *Theileria annae* (= *Babesia microti*-like) should be denominated *Babesia annae*.

Microscopic examination of blood smears has proven to be a low sensitivity method for detecting piroplasmid infection. New diagnostic procedures like the Polymerase Chain Reaction (PCR) and the Reverse Line Blotting assay (RLB) have been implemented to improve diagnosis. Up to date, no real-time PCR assays have been published for diagnosis of piroplasmosis in cats. By this reason, in this chapter is described a new method for diagnosis of feline piroplasmids and *Hepatozoon* sp. The procedure is based on quantitative PCR with FRET probes.

Chapter 3 - Hepatozoonosis is considered an emergent disease in felines. The genus *Hepatozoon* is a group of apicomplexan blood parasites found in a wide range of terrestrial vertebrates and from many hematofagous invertebrates. Infection of the vertebrate host usually occurs by ingestion of an infected invertebrate host. Although hepatozoonosis in felines is normally subclinical, pathogenic effects may be exacerbated in stressed or immunocompromised animals. Concomintant infections may also worsen hepatozoonosis prognosis.

The advent of genetic identification procedures has extended the spectrum of known feline hepatozoons. Several methods, such as standard PCR, PCR-RFLP or real-time PCR are now available for the diagnosis of hepatozoonosis. A new method for diagnosis of *H. felis* and *H. canis* in cats is described in this chapter. This procedure is complementary to the quantitative PCR method described in the preceding chapter.

Hepatozoon sp from cats has been confounded for a long time with *H. canis*, due to its morphological similarity with the species present in dogs. Interestingly, DNA sequencing has demonstrated that infection caused by *Hepatozoon* spp (and also by *H. canis*) is relatively frequent in felines. There is controversy about the classification of some hepatozoons, due to the lack of differential morphologic features. In this respect, it is formally proposed in this chapter to confirm the denomination of *Hepatozoon felis* for the two isolates of domestic cats previously sequenced in our laboratory and therefore, for those isolates genetically related to them. Such proposal is based on the different host range shown by *H. felis* and *H. canis*. This biological trait was confirmed by contingency table analysis. The idea is in agreement with the principle of priority in Zoological Taxonomy and moreover, is supported by the application of the molecular clock hypothesis to estimate divergence dates between species. In addition it is hypothesized that *Riphicephalus sanguineus* is one of the most likely vectors for *H. felis*, based on their common geographic distribution and their similar frequency in felines.

Although arthropods are the suspected definitive hosts for all feline hepatozoons, the application of vector control measures may have a limited efficacy for reducing disease levels in domestic cats and may prove very difficult to apply for wild felines. Due to the fact that hepatoozoonosis in cats is usually subclinical, there is no standard chemotherapeutic protocol available. There are, however, some reports of successful treatments with antibiotics and primaquine.

Chapter 4 - *Toxoplasma gondii* is a zoonotic intracellular protozoan parasite of worldwide distribution. The parasite is a main cause of abortion in small ruminants and humans when primary infection occurs during pregnancy. Felids, with particular importance of cats, are the definitive hosts, excreting oocysts in the faeces. Humans and virtually all warm-blooded species can be intermediate hosts and can become infected by ingestion of food and water contaminated with sporulated *T. gondii* oocysts, by consumption of tissue cysts in infected animal tissues, or congenitally. In the last decade, several studies of seroprevalence of *T. gondii* in numerous domestic and wild animal species (cats, dogs, pigs, wild felids, red deer and other cervids, wild boars, wild rabbits, wild birds and dolphins) have confirmed that *T. gondii* infection is widespread in Spain. Cats have a prominent role in the transmission of the infection. Recent studies in both domestic and wild cats from different areas of Spain have shown elevated seroprevalence levels, e.g: 45% of 220 domestic cats from Barcelona (Catalonia, North East Spain), and 32.3% of 585 domestic cats from Madrid and La Rioja (Central-North Spain). The highest seroprevalence has been observed in feral cats from Mallorca (Balearic Islands) (84.7% of 59 cats), being one of the highest reported worldwide in this species and the highest observed in Europe to date. High seroprevalence has also been observed in wild felids, such as Iberian lynx (81.5% of 27 animals, and 62.8% of 129 animals in two separate studies), and in European wildcats (50% of 6 animals). Furthermore, the presence of cats has been reported as one of the main factors for *T. gondii* seroprevalence in other domestic species, such as pigs and small ruminants in Spain. Seroprevalence levels in pregnant women in Spain have been reported to range from 18.8% in 2,929 pregnant women in Salamanca to 28.6% in 16,362 pregnant women in Barcelona. Since seropositive felids are likely to have already shed *T. gondii* oocysts in the environment, the high rate of *T. gondii* seroprevalence in cats can have important implications for conservation and animal infection and public health in Spain.

Chapter 5 - Acute or chronic uveitis is a common ophthalmic disorder that constitutes one of the main causes of irreversible blindness in domestic cats. Uveitis affects primarily the uveal tract, but other ocular critical structures can also be affected. In addition, inflammation of the posterior segment (choroid and retina) of the eye may provoke retinal detachment and degeneration, with consequent blindness. At present, treatment of cats with uveitis generally involves administration of corticosteroids, which is directed toward reduction of pain and lesions in ocular tissues. However, the immunosuppressive effect of corticosteroids may contribute to the development of the systemic disease, and their chronic use could result in corticosteroid-induced glaucoma. A novel method for experimental induction of uveitis in cats would greatly facilitate the understanding of the cellular events involved in ocular inflammation as well as the development of new treatment strategies. This chapter summarizes recent investigations showing that a single intravitreal injection of bacterial lipopolysaccharide (LPS) in cat eyes induces clinical, biochemical, functional, and histological changes that are consistent with the main features of naturally occurring feline uveitis, and that melatonin prevents clinical, biochemical, histological, ultrastructural, and functional consequences of experimental uveitis in cats.

In: Felines
Editors: R. García and R. Jiménez
ISBN 978-1-62257-005-8

Chapter 1

An Update on the Use of Type I Interferon for Treating Feline Retrovirosis

Ana Domenech[1], Nelida Gomez,[2] and Esperanza Gomez-Lucia[1,*]

[1]Facultad de Veterinaria, Universidad Complutense de Madrid, Spain
[2]Facultad de Ciencias Veterinarias,
Universidad de Buenos Aires, Argentina

Abstract

In this chapter we review the pathogenesis of the infections by feline immunodeficiency virus (FIV) and by feline leukemia virus (FeLV) as a way to understand the clinical signs. Even though there are commercial vaccines available, the reality is that these are two of the most prevalent infections in many areas of the world. As most retroviral infections, they are difficult to treat because the virus integrates into the feline DNA, behaving as a host gene. An alternative or complement to antiretroviral therapy is the use of type I interferons. Interferons are low molecular weight glycoproteins synthesized by several cellular types, mostly of the immune system (macrophages and T-helper cells).

* Corresponding Author: Esperanza Gomez-Lucia, DVM, PhD., Dept. Animal Health, Veterinary Sciences School, Complutense University, 28040 Madrid, Spain. E-mail: duato@vet.ucm.es.

They are found naturally in the serum of animals infected by viruses and represent one of the components of the intrinsic immunity. Due to their antiviral activity, their potential to boost the natural immune response, and to the clinical results obtained with HIV, interferons have been proposed for treating feline retrovirosis. However, their efficacy in this respect is controversial. Though long-term studies are lacking, it seems that in the short-term the clinical condition of cats improves transiently with type I interferon (IFN) treatment, but this is not associated with a decrease in viral or proviral loads. Studies on the in vitro effect of IFN to clarify whether or not IFN affects the retroviral cycle suggest that this chemokine may decrease the number of infective viral particles, but does not act directly on viral expression but rather on morphogenesis and budding. Reports on the lack of specific sequences in the control regions of the proviral genome (LTRs) of FeLV and FIV agree with these in vitro results. Most studies have been carried out on infected cats with no tumors, and the possible use of interferons for treating the different tumors which may accompany feline retroviral infections is still far from confirmed. Thus, more studies are still needed to understand better the effects of interferon treatment in feline retrovirosis.

INTRODUCTION

Feline leukemia (FeLV) and feline immunodeficiency virus (FIV) are two important pathogens of cats which induce chronic infections which eventually lead to the progressive weakening of cats and the presence of multiple clinical signs. Both infections are characterized by immune suppression in advanced stages that allow secondary infections to establish and may lead to the death of the animal. There is no effective method to eliminate the virus in infected animal and treatment is usually based on supportive and symptomatic therapy [Hartmann, 2006; Sellon and Hartmann, 2006; Dunham and Graham, 2008]. For some time type I interferons have been proposed as an alternative for treatment, as they might have an in vivo antiviral effect, which would add to their immunomodulator and antiproliferative (antitumor) effect, but there is no conclusive data to date. The aim of the present review was to provide an update on the data on type I interferon therapy for treating feline retrovirosis, evaluating its advantages and disadvantages.

GENOMIC CHARACTERISTICS OF FELINE RETROVIRUSES

Feline immunodeficiency virus (FIV) and feline leukemia virus (FeLV) are two of the most prevalent causes of feline infections worldwide. FeLV, a gammaretrovirus, has a simple genome consisting only of the genes necessary for replication and particle formation (*gag, pol* and *env*). For this reason, it can only replicate in actively dividing cells, such as those in the bone marrow or lining the small intestine [Dunham and Graham, 2008]. On the other hand, FIV is a lentivirus, proposed as the best animal model for the study of human immunodeficiency virus (HIV). As such, its genome, besides the above mentioned genes, has other genes which encode accessory proteins that enable the virus to regulate the replication cycle more efficiently, thus productively infecting non-dividing cells [Dunham and Graham, 2008].

The proviral genome of all retroviruses is flanked by sequences known as long terminal repeats or LTRs. The 5' and 3' LTRs are identical. Three different regions can be identified in them: U3, R and U5. Transcription begins in the first nucleotide of R. Several promoters and enhancers have been identified in U3, some of which may react with cellular factors, such as hormones [Tejerizo et al., 2005; Domenech et al., 2009] or cytokines [Robert-Tissot et al., 2011], which thus govern viral expression.

TRANSMISSION

Saliva is the main source of infection for feline leukemia virus, as viremic cats with or without clinical signs shed constantly viral particles through it [Hartmann, 2006]. Mutual grooming or sharing food or water bowls are risk factors for FeLV transmission and for this reason feline leukemia is known as the "friendly cats' disease". Contact needs to be close and extended, as the virus is very labile in the environment and sensitive to ultraviolet light, heat, detergents and desiccation. The importance of other secretions on transmission is notably lower [Hartmann, 2006].

On the other hand, the main transmission route of FIV is the inoculation of blood and/or saliva, which usually happens through bites and wounds produced by aggressions. For this reason, feline immunodeficiency is the "unfriendly cats' disease" [Arjona et al., 2000; Dunham and Graham, 2008; Hosie et al., 2009; Lutz et al., 2009]. Queens may transmit FIV to the kittens during pregnancy or nursing [Burkhard and Dean, 2003], and not all the kittens from a litter may acquire the infection [Sellon and Hartmann, 2006]. Other routes, especially those involving mucosa, are possible [Burkhard and Dean, 2003].

These different transmission mechanisms determine the beginning of the pathogenesis. If FeLV is not cleared early from the body by an initial strong immune response, it produces a persistent infection. Characteristically, the virus remains latent in cells of the bone marrow, but eventually animals die as a result of it. On the contrary, when FIV is inoculated into a bitten cat it replicates in monocytes, macrophages and lymphocytes, and spreads throughout the body. The decrease in lymphocytes and macrophages targeted by FIV induces a progressive breakdown of the immune system [Bendinelli et al., 1995; Dunham and Graham, 2008; Hosie et al., 2009], which may lead to death.

PATHOGENESIS

Both feline immunodeficiency and feline leukemia have aspects in common: they are persistent infections affecting mainly lymphocytes and other immune cells, so they are accompanied by alterations or imbalances in the immune system and its function, and are characterized by a variety of signs that are clinically difficult to differentiate. However, their pathogenic mechanisms are quite different and not completely understood. The main difference is that FIV has a direct pathogenic action, causing death by apoptosis of its main target-cells, $CD4^+$ T-cells, and therefore it does not depend on the immune status of the animal. On the contrary, FeLV infection is greatly influenced by the ability of the cat to develop an immune response [reviewed by Domenech et al., 2007]. Another big difference is that recovery from FIV infection has not been reported, while the majority of cats exposed to FeLV are able to clear the infection if they develop an initial strong immune response [reviewed by Dunham and Graham, 2008].

As mentioned above, FIV usually is transmitted through bites. After an initial replication in the regional lymph nodes and other lymphoid organs, a short initial viremia occurs in which virus is spread throughout the body, infecting lymphocytes, monocytes and macrophages of the bone marrow, intestinal tract, lung, kidney and brain. A few weeks later, the cat develops a cellular and humoral immune response which, however, fails to completely eliminate the virus. The infection progresses to a chronic asymptomatic period, which corresponds to the phase of clinical latency, with low levels of viral expression, and which may be lifelong lasting in some animals. During this asymptomatic phase the immune system weakens gradually, with a progressive decrease of $CD4^+$ T-cells (due to mechanisms of apoptosis mediated by FIV) and of the CD4/CD8 ratio, along with a decrease in the functionality of infected lymphocytes. In some infected cats, often many years after infection and coinciding with very low values of CD4/CD8, plasma viremia increases, leading to the terminal clinical disease, the so-called "feline acquired immunodeficiency syndrome" or FAIDS, due to its resemblance to human AIDS. At this stage, opportunistic infections occur, against which the cat cannot develop an adequate immune response. Finally the cats die from a chronic syndrome of wasting, neurological disease, neoplasia or systemic opportunistic infections [Reviewed in Domenech et al., 2007; Dunham and Graham, 2008; Hartmann 2011].

The outcome of FeLV infection is very different in each cat, as it depends on factors related to both the virus (subgroup of FeLV, duration of exposure and infective viral load) and the host (age and immune status) [reviewed in Domenech et al., 2007; Dunham and Graham, 2008; Hartmann, 2011]. The animal usually gets infected by the oronasal route, and the virus replicates in the local oropharyngeal lymphoid tissue. Depending on the immune response the infection progresses differently. Most cats (60%) have an effective immune response, mainly cell mediated (CTLs) but also antibody mediated, and the virus is completely cleared from the animal at a very early stage ("abortive infection") or after a transitional phase of viremia in which the virus infects lymphocytes and monocytes, and is spread to different tissues (spleen, thymus, salivary glands, lymph nodes) ("regressive infection"). These animals are protected against future exposures to the virus. However, in 30% of cats where the immune response is not effective in eliminating the virus and viremia persists more than three weeks, FeLV infects hematopoietic stem cells in the bone marrow, producing a massive replication of the virus and the development of persistent viremia (both of free virus and of infected cells), which gives rise to "persistent or progressive infection". In addition, hematopoietic alterations (particularly cytopenias) are produced, caused by the suppression of the bone marrow as a result of the infection [Reviewed by Hartmann, 2011]. These animals often die in two or three years from a disease related to FeLV infection. In an unknown percentage of cats, the bone marrow becomes infected, but does not release virus or infected cells, a condition known as "latent infection", and animals remain apparently healthy until the provirus is reactivated by an immunosuppressor or treatment. Once the bone marrow becomes infected, the virus cannot be completely cleared, since the cell progeny also carries the provirus. Very young cats are more likely to develop a progressive infection, probably due to their immature immune system, while older cats tend to develop a regressive or abortive infection. Atypical infections in which the virus is located in organs, as urine bladder, eyes or mammary gland, partially hidden from an effective immune response, have been described in a very low number of animals. Certain immunosuppressive FeLV strains (FeLV-T), with a strong tropism for T-cells, have also been reported to induce an acquired immunodeficiency syndrome similar to the one produced by FIV [Reviewed in Hartmann, 2011].

Since these retroviral infections in cats induce deep deregulation of the functions of the immune system, both infections are associated with alterations of the humoral immune response. In most cats infected with FIV, there is a marked hypergammaglobulinemia, primarily due to the polyclonal stimulation of B-cells which give rise to an increase of IgG not specific to FIV [Flynn et al., 1994; reviewed in Doménech et al., 2007; Gleich and Hartman, 2009; Hartmann, 2011]. The excess of circulating immune complexes described in both infections could contribute to the immune mediated lesions and symptoms associated with the infection by any of these two viruses (see below) [reviewed in Domenech et al., 2007; Hartmann, 2011].

Changes in the expression of the synthesis of cytokines or of their cellular receptors have been described in feline retrovirosis, especially in FIV infections. Reported alterations in the levels of IFNγ, TNF-α, IL-4, IL-6, and altered IL-10/IL-12 ratio could influence the shift from type Th1 response (which contains disease progression) to type Th2 (which allows disease progression) during FIV infection [Levy et al., 1998; Levy et al., 2004; Lehman et al., 2009].

The presence of tumors has been described in both infections. FeLV is an oncogenic retrovirus that causes different tumors, mainly lymphomas and leukemias (see below). The main oncogenic mechanism is by insertion of the viral genome in the vicinity of a cellular oncogene, producing its activation or over-expression and the uncontrolled cellular proliferation. FIV infection also is associated with a greater presence of tumors in cats, mainly B-cell lymphomas, but also leukemia and other tumors [Callanan et al., 1996; Beatty et al., 1998; Gabor et al., 2001; Sellon and Hartmann, 2006; Hartmann, 2011]. The provirus is only detected occasionally in tumor cells [Beatty et al., 1998; Sellon and Hartmann, 2006; reviewed in Hartmann, 2011], and it seems that the virus is not involved directly in their development, but through the decrease of cell-mediated antitumor immunity or through the chronic hyperplasia of B-cells [Beatty et al., 1998].

Finally, feline retrovirosis are often accompanied by neurological disorders which originate from the viral infection of brain endothelial cells, microglia and astrocytes during the viremic phase [reviewed in Fletcher et al., 2011]. The mechanisms involved in the neuropathogenesis are not fully known, but in FIV infection they include the apoptosis of neurons (although they do not become infected), release of toxic products by infected microglia, disorders of cell metabolism of the astrocytes or cytokine production as a result of the infection. $CD8^+$ T-cells also seem to be involved [reviewed in Domenech et al., 2011; reviewed in Fletcher et al., 2011].

It seems clear that neuropathogenesis is linked to certain FIV strains, probably associated with their envelope proteins [Power et al., 2004; reviewed in Fletcher et al., 2011]. A direct neurotoxic mechanism has been proposed for FeLV through its envelope glycoproteins that are able to increase the concentrations of intracellular calcium, leading to neuronal death, as described in certain HIV strains [reviewed in Domenech et al., 2007].

Clinical Signs

Clinically, feline immunodeficiency is characterized by its chronic nature. A FIV-infected cat may remain asymptomatic for many years. Three stages may be recognized: acute phase, clinically asymptomatic carrier phase, and symptomatic stage (FAIDS phase). The acute phase, which generally lasts 4 to 6 weeks but can persist up to 1 year, is characterized by hyperthermia, generalized lymphadenopathy, diarrhea, depression, skin lesions, intraocular inflammation, variable neurologic signs, etc. [Caney, 2000; Hosmalin and Lebon, 2006; Lawson et al., 2009]. During the asymptomatic carrier phase, which can last from months to many years, gradual alterations of the immune system (decrease of the CD4/CD8 ratio and hypergammaglobulinemia) occur. Non severe clinical signs, such as relapsing fever, leucopenia, lymphadenopathy, anorexia, intermittent weight loss, etc. may be observed in this stage. When the triad lymphadenopathy-gingivitis-hypergammaglobulinemia is observed, feline immune deficiency virus infection should be suspected [Gómez, 2002; Lawson et al., 2009], and, if the cat is positive, antiretroviral therapy should be started. Several immune-mediated clinical signs may be detected (uveitis, glomerulonephritis, cholangitis, etc.) and very few are due to direct viral action [Lawson et al., 2009; Hartmann, 2011].

During the symptomatic disease phase the lymphocyte depletion is much more pronounced and chronic organ and tissue infections are observed, such as gingivo-stomatitis, respiratory diseases, chronic infections of the skin, the nervous system, the kidneys, or the eyes, etc. When the CD4/CD8 ratio decreases further, acquired immunodeficiency, characterized by the presence of opportunistic infections and tumors, such as squamous cell carcinoma and lymphoma, is produced. This stage is termed FAIDS, due to its similarity to humans AIDS [Gómez, 2002; Little et al., 2009; Hartmann, 2011]. It usually ends with the death of the animal as a result of the opportunistic infections.

Though FeLV infection also induces immune suppression, the clinical signs induced are more specific than those observed in the course of FIV infection and the evolution is faster, especially in persistently infected cats, which usually die in 1-3 years [Hofmann-Lehmann et al., 2007; Hartmann, 2011]. FeLV-infected cats may show neoplastic and/or non-neoplastic disease. As mentioned above, FeLV is an oncogenic virus which may induce several types of neoplasias, of which lymphoma and leukemia are the most prevalent. RNA may be detected in the neoplastic lesions, even when cats with neoplasias are often serologically negative to p27, *i.e.* many of them are non-viremic. Besides lymphoma (located in the mediastinum, the mesenterium or extra-nodal), and lymphoblastic leukemia, cats may show myeloproliferative neoplasias, fibrosarcoma, osteocondroma and neuroblastoma [Pedretti et al., 2006; Hofmann-Lehmann et al., 2007; Hartmann, 2011].

As regards non-neoplastic alterations, anemia is frequently observed in viremic animals and is predictive of imminent death. It may be non-regenerative (as a result of myelosuppression, myelodestruction, or myeloproliferative diseases) or hemolytic (immune-mediated or secondary to *Mycoplasma haemofelis* infection) [Pedretti et al., 2006;; Hofmann-Lehmann et al., 2007; Hartmann, 2011]. Immune suppression, associated with the viral infection of platelets and neutrophils, which induces alterations in chemotaxis and phagocytosis, predisposes animals to secondary infections, such as stomatitis, abscesses, pyothorax, dermatitis, feline infectious peritonitis, toxoplasmosis or cryptococcosis [Hartmann, 2011]. FeLV infection may also cause infertility and abortions, and newborn kittens are frequently lethargic. Some immune-mediated alterations observed in FeLV positive cats are autoimmune glomerulonefritis, uveitis and neutrophilic polyarthritis [Hofmann-Lehmann et al., 2007; Hartmann, 2011]. Some cats have anisocoria syndrome, resulting from the effect of the virus on the neurovegetative system, which may also produce urinary incontinency [Hofmann-Lehmann et al., 2007].

VACCINES

There are three types of vaccines against feline leukemia: based on inactivated complete virions, on protein subunits or on recombinant viruses [*American Association of Feline Practitioners*, 2006; Hofmann-Lehmann et al., 2007]. The first two types have been linked to the development of post-inoculation sarcoma in several vaccinated cats, probably due to the need of using adjuvants, and have discouraged owners from vaccinating their cats [*American Association of Feline Practitioners*, 2006]. Nowadays, recombinant vaccines are preferably used, because they are considered to be very immunogenic and with low probability of side effects [*American Association of Feline Practitioners*, 2006; Hosie et al., 2009]. Feline leukemia vaccination is highly recommended for kittens and revaccination is advised depending on the lifestyle of the cat [*American Association of Feline Practitioners*, 2006].

As regards FIV, several vaccines have been studied experimentally, which are based on inactivated virus or on inactivated infected cells, recombinant viruses (canarypox based vaccines) and DNA vaccines [Willett and Jarrett, 1995; Uckun et al., 2003; *American Association of Feline Practitioners*, 2006; Hosie et al., 2009]. The vaccine that has proved most effective against the strains detected in USA and Canada is based on inactivated whole virus and it is sold in USA, Canada and New Zealand [*American Association of Feline Practitioners*, 2006]. The effectiveness of this vaccine has not been tested for other strains of FIV but some studies demonstrate that it is not effective against the existing UK strains [Hosie et al., 2009]. Vaccination induces the formation of antibodies specific for FIV for at least 1 year which cannot be differentiated from those produced during the disease. As this interferes with the serological diagnosis, it is not licensed for use in the European Union.

TREATMENT

Regardless of the availability of vaccines, prevalence of feline leukemia and of feline immunodeficiency is very high in some countries and infected cats still require to be treated. The emergence of HIV has prompted the study of different control strategies for other retroviral infections. Besides, FIV-infection of cats is considered as the best animal model for the study of the human virus and drugs and protocols are tested in cats to ensure proper antiretroviral efficiency and low toxicity in humans. The best results are obtained by HAART (Highly Active AntiRetroviral Therapy), which combines drugs that act at critical points of the retroviral cycle such as viral entry, reverse transcription of the genome, integration of viral genome in the cellular DNA, and the assembly of the viral proteins [Arai et al., 2002; Bisset, 2002; Uckun et al., 2003]. Drugs combinations decrease the emergence of viral mutations usually observed in single drug therapy [Arai et al., 2002; Uckun et al., 2003; Gisbert and Gómez, 2010]. Reverse transcriptase inhibitors include nucleoside analogues, nucleotides analogues, and non-nucleosidic analogues. They avoid the conversion of viral RNA to viral DNA, arresting the cell cycle and rendering viral RNA susceptible to degradation in the cytoplasm [Arai et al 2002; Bisset, 2002; reviewed in Collado et al., 2007b; Gisbert and Gómez, 2010]. The zidovudine (ZDV or AZT) (nucleoside analogue) is the reverse transcriptase inhibitor most widely studied in cats. It produces the decrease of viral titers in plasma and clinical and quality of life improvement, but it may induce anemia, vomiting and anorexia. Lamivudine (3TC) is another nucleoside analogue, which may revert the undesirable mutations produced by ZDV, allowing a longer treatment with it [Arai et al., 2002; Bisset, 2002]. However, results suggest that it is effective only when administered previous to infection, but not in cats with chronic infection [Arai et al 2002; reviewed in Collado et al., 2007]. Didanosine [DDI], well tolerated in humans, produces in cats severe and irreversible secondary effects such as anemia, thrombocytopenia and neurologic alterations [Bisset, 2002]. Stampidine, another reverse transcriptase inhibitor, has a weaker effect than ZDV, but when both drugs are combined they act synergically [Uckun et al., 2003; Collado et al., 2007a].

Recent researches introduce protease inhibitors (such as TL-3) in the drug cocktails, which may avoid neurologic clinical signs [Torbett et al., 2003; Uckun et al., 2003]. However, other protease inhibitors, such as sequinavir, indinavir, ritonavir and nelfinavir, have not been shown to effectively inhibit FIV replication. Other drugs are the inhibitors of viral integrases, which are presently studied in humans and which show promising results in cats [de Rozieres et al., 2004; Savarino et al., 2007].

Valproic acid, an anticonvulsant drug used in humans, has been shown to be an inhibitor of the enzyme histone deacetylase, responsible for HIV to stay silently incorporated into cellular DNA in the latent phase of infection and resistant to conventional therapy [reviewed in Collado et al., 2007a]. Valproic acid causes the activation of the provirus, which can thus be neutralized by other drugs [Dicenzo et al., 2004; reviewed in Collado et al., 2007a]. However, our studies with FIV-infected cats do not confirm this beneficial effect [Gisbert and Gómez, 2010].

It should be remembered that, though feline retrovirosis resemble human HIV infection, differences exist both between HIV, FeLV and FIV and between humans and cats. There are drugs which are safe and/or effective in humans, but are toxic and/or ineffective in cats, such as DDI [Arai et al., 2002], and vice versa, such as TL-3 [Torbett et al., 2003; Uckun et al., 2003]. Therapies employed in humans should not be extrapolated blindly to felines without first carefully testing their effectiveness and their undesirable effects.

Immune Modulators

Non-specific immune modulators have also been proposed for treating these feline retrovirosis, as both FIV and FeLV produce a marked deterioration of the immune system. However, long-term field studies to determine the benefits and disadvantages of immuno-stimulants are lacking [Gómez, 2000; Lawson et al., 2009; Domenech et al., 2011]. Some of the immune modulators examined so far for treating FeLV-infected cats and to a lesser degree FIV-infected cats are *Propionibacterium acnes*, Acemannan (a polysaccharide derived from the plant Aloe Vera), bovine lactoferrin (an iron regulatory protein present in milk, which has anti-inflammatory effects), protein A from *Staphylococcus aureus*, and the attenuated *Mycobacterium bovis* known as Calmette-Guerin bacillus (BCG). These and other non-specific immune stimulants activate macrophages and cellular immunity and have some antiviral activity by inhibiting glucose metabolism [Fernandez et al., 2011].

When considering treating a retrovirus-infected cat differences between FIV and FeLV infections should be taken into account. Generally, when FeLV-infection is diagnosed (persistent viremia) the patient's immune system is greatly impaired and life expectancy is short. For this reason it might be wiser to use specific stimulants and to start therapy as soon as possible.

In FIV-infection, the disease is chronic and the beginning of treatment depends on the clinical status of the patient, viral load and the CD4/CD8 ratio [Gómez et al., 2002; Riondato et al., 2003; De Mari et al., 2004; Phillips et al., 2005; Collado et al., 2007a]. When the patient is still in the asymptomatic stage, the immune system is still fully functional and organs are not damaged, so it tolerates therapy better. Ideally, treatment should begin when cats are still in this stage but about to enter the final phase of the disease, when severe complications or secondary infections due to opportunistic agents are seen

[Gómez, 2000; Little et al., 2009; Gisbert and Gómez, 2010; Hartmann, 2011]. During FAIDS, response to therapy is not as good as in the asymptomatic stage and treatment must be frequently suspended due to anemia. Complications and secondary infections should be treated first and then move on to antiretroviral therapy.

In Vivo Studies with Type I Interferons

Type I interferon (IFN) is the immune modulator which has been most extensively studied for the treatment of retroviral infections. The possible uses of interferon (IFN) in FIV- or FeLV-infected cats are based on the knowledge of its effects in HIV-infected humans and, although there are many similarities, it should be remembered that two different species and three different viruses are considered. For example, IFN therapy induces signs of depression in humans, while it has a euphoric effect in cats. Regardless of the dose, IFN therapy improves the mood of the treated cat, which is much more active and with increased appetite [personal observation, NG].

The use of type I IFN for treating retroviral diseases is based on the observation that asymptomatic non-progressor HIV-infected patients have increased IFN-α concentrations [Soumelis et al., 2001], which correlate with higher number of $CD4^+$ T-cells, low HIV titers and absence of opportunistic infections [Feldman et al., 2001; Soumelis et al., 2001; Finke et al., 2004]. In addition, type I IFNs increase the expression of class I MHC molecules in all cells, favoring the removal of infected cells [Andrieu et al., 2001].

The use of IFN for treating HIV infections is still a controversial issue [Fernandez et al., 2011]. Several explanations are provided as to whether IFN treatment can be detrimental or beneficial for HIV patients. The current trend is geared to the use of specific stimulants of the immune system and not so much to the nonspecific stimulants. Some of the reasons for not using IFN therapy in HIV-infected patients are: (a) type I IFN induces cytotoxic activity of the antigen presenting cells, which may contribute to decreasing T-cell numbers, either by apoptosis of non-infected $CD4^+$ T-cells, or inhibiting the proliferation of T-cells and cytokine production [Hosmalin and Lebon, 2006]; (b) IFN stimulates tetherin, a protein that induces new virions to adhere to the cell surface, preventing their spread to other cells, but which may stimulate the development of viral variants able to pass by cell-to-cell contact and affect disease progression [Fernandez et al., 2011]; (c) the blood levels of IFN are very low or undetectable in healthy individuals, but high in the acute infection and in the final stage of HIV, so the administration of exogenous IFN, when the endogenous levels are already high, is debatable [Fernandez et al., 2011]. Accordingly, some researchers recommend treating HIV patients with IFN only when it is associated with other drugs and when infected with Hepatitis C virus and Kaposi's sarcoma herpesvirus [Fernandez et al., 2011].

For some time IFNs have been proposed as an alternative in the treatment of feline retrovirosis. Several approaches following different protocols have been studied. Most studies conclude that IFN-treated cats improve clinically, though few effects are seen on the viral loads. Two molecules of type I interferons are presently used, the recombinant human alpha interferon (rHuIFN-α) and the recombinant feline omega interferon (rFeIFN-ω). Initial trials used rHuIFN-α, the only type I IFN available in the market at the moment. Treatment with high-doses (10^6 U rHuIFN-α /kg), similar to those used in therapies against HIV infections, were suspected of inducing anti-interferon antibodies. Currently, lower doses are used which do not induce this humoral immune response [Hosie et al., 2009].

Initial reports showed that treatment with rHuIFN-α decreased the mortality of cats experimentally infected with FeLV [Cummins et al., 1988] or with natural infection [Cotter, 1979]. It was later described that the clinical and analytical status of FeLV-infected cats improved with treatment with rHuIFN-α [Steed, 1987; Weiss et al., 1991], and antigenemia decreased [Zeidner et al., 1990]. However, improvement in these parameters was not observed by others [McCaw et al., 2001]. Orally administered rHuIFN-α at low doses (30-60 IU/SID) for alternate weeks has proven to be effective in non-neoplastic phases of FeLV infections [Arai et al., 2002; Gómez et al., 2002; Torbett et al., 2003], improving the general condition and the survival rate of infected cats [Uckun et al., 2003].

Studies on treatment of cats infected with FIV are scarce. A controlled study in the UK with 5 sick FIV-infected cats, treated with rHuIFN-α, followed over 1.5 years did not detect significant differences in viral loads and CD4/CD8 ratio [Caney, 2000]. In another study, 30 FIV-infected cats with advanced clinical disease were treated with 10 IU/Kg of rHuIFN-α and 6 with placebo. Interferon treatment improved the general condition of cats, increased $CD4^+$ T-cell counts, and produced a slow but progressive increase of $CD8^+$ T-cell counts. However, no significant differences were observed in the viral and proviral loads. A rapid regression of the immunopathological signs was observed, which suggests that IFN-α affects the cytokines circuits without influencing viral persistence [Pedretti et al., 2006].

With the marketing of recombinant feline IFN-ω (rFeIFN-ω) most of the studies on feline medicine focused on its use. Cats infected with FeLV or coinfected with FIV and FeLV and treated for 5 days with 1MU of rFeIFN-ω /kg/day had a decreased mortality rate and an improvement in the clinical condition and the counts of white blood cells when compared to placebo-treated cats, although no positive evolution of the red blood cells series was observed [de Mari et al., 2004].

The laboratory which commercializes rFeIFN-ω (Virbagen®, Virbac, France) indicates that this cytokine has immune modulatory and antiviral properties. Although there is no conclusive data for the moment, it has been proposed that the beneficial effects observed when treating with rFeIFN-ω could be due to the improvement in the immune status of the animals rather than to a direct antiviral effect [Riondato et al., 2003; Collado et al., 2006; Collado et al., 2007a; Phillips et al., 2005]. A recent study with FIV- and FeLV- infected cats showed that IFN-treated cats underwent an evident improvement of the clinical condition, as well as of the hemogram parameters, without improvement of the viral- or immunological-related parameters. Thus, these results suggested that the overall effect of rFeIFN-ω could be on the innate immune system, rather than on the adaptive [Domenech et al., 2011]. rFeIFN-ω has been used for treating other feline viral infections, including herpesvirus and calicivirus, but its action on them is not well demonstrated either [Hosie et al., 2009].

Regardless of these positive results on the clinical condition of the IFN-treated cats observed in several studies, the European Advisory board on cats' diseases [Hosie et al., 2009] considers that there is no conclusive evidence showing that immune modulators and IFNs have beneficial effects on the health and survival of symptomatic and asymptomatic FIV-patients. Moreover, a non-specific stimulation of the immune system may be contraindicated because it could lead to a rise of viral replication produced by the activation of lymphocytes and macrophages infected in latent form and therefore accelerate disease progression [Hosie et al., 2009]. In addition, the benefits of IFN-treatment have not been unequivocally shown in long-term studies, and the effect may be transient.

In summary, type I interferon stands as a good alternative for the therapy of feline retrovirosis as most researchers agree that the clinical status of treated cats improves notoriously, though the effect on the viral load may be uncertain.

In Vitro Studies on the Effect of Type I Interferons

Few studies have been carried out to address the question as to whether or not type I interferons affect the feline retrovirus cell cycle. In the case of HIV-1, it has been reported that type I IFNs inhibit the *in vitro* replication of this virus [Bednarik et al., 1989; Poli et al., 1989; Gendelman et al., 1990], especially when cells are pretreated with interferon, and this effect is less evident when treated post-infection [Meylan et al., 1993; Agy et al., 1995]. The expression of Mx and OAS genes induced by type I IFNs (see below) inhibit the formation of viral particles and reduces reverse transcriptase (RT) activity [Baca-Regen et al., 1994] and of p24 protein [Smith et al., 1991] of HIV-1. In addition, virions which bud are 10-1000 times less infective tan those from non-treated cells [Hansen et al., 1992; Dianzani et al., 1998]. On the other hand, other studies have shown that the amounts of HIV-1 mRNAs are not altered by treatment with interferon [Coccia et al., 1994].

In the case of FeLV, results suggest that IFNs do not apparently affect viral protein synthesis, at least of p27, since its concentration in the supernatant was unaffected by different concentrations of either rHuIFN-α or rFeIFN-ω [Collado et al., 2007b]. However, reverse transcriptase activity, a measurement of infective viral particles, decreased with increasing concentrations of IFNs and longer treatment times. All this suggested that interferon may interfere with the final stages of the cycle, the processing of viral proteins and their assembly into virions [Collado et al., 2007b], similar to what has been described for HTLV-I [Feng et al., 2003]. On the other hand, other studies seemed to demonstrate that interferon may efficiently inhibit the early stages of replication, decreasing the levels of integrated provirus, viral RNA and expressed proteins, at least in the case of HIV-1 [Shirazi and Pitha, 1992]. It was also shown that IFNs induced a decrease in viability of FeLV-infected cells, which seemed to die rather by apoptosis than by necrosis [Collado et al., 2007b]. This would limit the spread of the infection to other cells, and to the selective death of infective cells [Collado et al., 2007b].

Molecular Mechanisms for the Induction of Type I Interferon

Interferons are inducible proteins, *i.e.* they are not present constitutively but synthesis is prompted by a stimulus. Induction of type I IFNs is regulated by two major and distinct-cellular signal transduction pathways, to sense viruses and activate cellular type I IFN genes [reviewed in Haller et al., 2006]. Most-cells, including fibroblasts, hepatocytes and conventional dendritic cells (cDCs), use the so-called classical pathway, initially secreting IFN-β, which triggers the expression of IRF-7, proposed as the main regulator of the expression of type I IFN genes [Honda et al., 2005]. IRF-7 and IRF-3 cooperate to initiate the synthesis of IFN-α [Marié et al., 1998; Haller et al., 2006]. In contrast, plasmacytoid dendritic cells (pDCs) constitutively express IRF-7, and thus they can by-pass the initial activation, predominantly secreting high levels of IFN-α. They represent the so-called natural interferon producing cells of the body [Haller et al., 2006].

When IFN is synthesized, it is secreted to the extracellular environment. In order for cells to acquire the “anti-viral state”, IFN needs to enter them after binding the appropriate receptors. The specific receptor for type I IFN, called IFNAR, is formed by two subunits: IFNAR-1 and -2. Feline IFNAR-2 has recently been molecularly characterized and found to have 61% homology at the amino acid level with human IFNAR-2 and a structure similar to other IFN receptors [Xue et al., 2010]. When type I IFNs bind IFNAR, both subunits heterodimerize [Mogensen et al., 1999], phosphorylating the so-called JAK-STAT pathway. The resulting trimer translocates to the nucleus and binds to DNA regulatory sequences called ISREs (IFN-Stimulated Response Elements) in the promoter region of the Interferon Stimulated Genes (ISGs). ISREs have the consensus sequence AGTTTCNNTTTCNP. Specialized proteins serve as negative regulators and inhibitors of the JAK-STAT pathway.

Type I IFNs activate the expression of several hundred ISGs, the products of some of which have antiviral activity. Proteins involved in the antiviral response may act through three main pathways: Protein kinase R (PKR), 2’,5’-oligoadenylate synthetase (OAS) and Mx proteins [Haller et al., 2006]. Another pathway which has been paid attention in relationship with FIV infection is ISG15 [Robert-Tissot et al., 2011]. The addition of this protein is called ISGylation and it produces a ubiquitin-like posttranslational modification. The feline ISG15 gene has been identified and found that the capsid protein of FIV was ISGylated *in vitro* by treatment with feline IFN-ω, a mechanism which might block the virus [Shimode et al., 2012]. Microarray technology is helping identify other genes whose expression is modified by interferon.

In addition to activation of metabolic pathways induced by interferons, type I IFNs also induce other molecules which play a role in anti-viral defense. One of such molecules, tetherin (also called BST-2 or CD317), has been studied in cats and has been shown to potently inhibit FIV particle release. It is similar to the human counterpart in amino acid sequence, protein topology and antiviral activity. However, FIV may lack a functional tetherin antagonist since cell-to-cell spread remains efficient in the presence of abundant viral receptors and tetherin upregulation may even enhance syncytium formation [Dietrich et al., 2011].

Direct Effects of Type I Interferons at the Molecular Level: ISREs

As mentioned previously, ISREs are DNA sequences which bind the well-conserved N terminal region of the Interferon Regulatory Factors (IRFs). ISREs are present on the promoters of the target genes, *i.e.* the IFN-α or IFN-β genes and the IFN stimulated genes (ISG). By this mechanism, IRFs regulate the expression of genes stimulated by IFN, either activating or repressing gene transcription, depending on the target gene.

The expression of retrovirus genes is under the control of the LTR which contains sites which bind both viral and host transcription factors. The presence of sequences able to combine IRFs in this region could affect viral expression in the presence of interferon. In HIV-1 a sequence (TTGAAAGCGAAAGGGAAACC) homologous to ISRE has been identified in the 5' LTR downstream of the HIV-1 transcription start site. This sequence has been shown to be a binding site for several IRFs (IRF-1, IRF-2, IRF-3) and its deletion results in impaired LTR promoter activity and altered synthesis of viral RNA and proteins [reviewed in Gomez-Lucia et al., 2009].

No ISRE sequences have been identified in FIV isolated from infected cats but it has been described that the LTR of different FIVpco strains (isolated form cougars) have recognition sites for different interferon response factors: IRF-1 and IRF-3 [Poss and Ross, 2008]. IRF-1 enhances transcription of genes regulated by IFN-α and the site for this factor (99.3% similar to the site identified in humans) is located at position 16-24 in the FIVpco LTR.

IRF-3 is induced by virus infection and regulated by phosphorylation. However, an absence of IRF-3 site has been described in a FIVpco strain that has been maintained in tissue culture for over 15 years. These two IRF sites may enhance virus replication consequent to an innate interferon response [Poss and Ross 2008 and references therein]. Recently, some putative binding sites for Transcription Factor Binding Sites (TFBS) involved in IFN gene regulation system have been identified in FIV_{GL8} strain (isolated from cat) using the software Interferome (version 2009.3, Transfac Professional) [S. Samarijiwa, personal communication]. A matched sequence (TBP) (GATAAATGGAAACAG) was located in both LTRs of FIV_{GL8}, at positions 71 and 9184, and others in different regions of the genome at positions 2967 (*pol*, GGGGATTTGAAACTC), 7384 (*env*, AAGTTTCATTGTCAG) and 7489 (*env*, GAAAAAGTGAAAGTA). However, it is difficult to link these sequences outside the LTR region with the initiation of transcription of the viral genome.

In FeLV, the sequence GGTTTCATTTTCG, matching the consensus NAGTTTCNNTTTCNC/T [Williams, 1991], was found at nucleotide 798 in FeLV-A Rickard strain (FeLV GenBank Accession Number AF052723), in a region just upstream from that which encodes the Gag-Pol precursor polyprotein gPr80; as before, it is difficult to ascribe for a sequence outside the control LTR region a role in transcription [Collado et al., 2007b]. Thus, the results of the analysis of the genome of FeLV agree with the observations described above (Section In vitro studies on the effect of Type I Interferons) that IFN does not affect the synthesis of viral proteins.

Antitumor Activity of Type I Interferon

IFN-α has clear anti-proliferative effects in cells in culture, which has led to investigate its possible in vivo antineoplastic properties as a potential anti-cancer drug [revised in Brassard et al., 2002; Bracarda et al., 2010; Kiladjian et al., 2011], an ability which would be interesting in feline retrovirosis associated tumors. In addition, type I IFN may act also indirectly through the regulation of certain immune mechanisms (CTLs and cytokines, innate cells). Though in several animal models tumors are seen to shrink or even remit following IFN administration, the mechanisms involved are not clear [reviewed in Bracarda et al., 2010]. The doses and duration of optimal treatment are also unknown, although it appears that the anti-tumor activity is associated with high repeated doses, which poses problems of toxicity and adverse effects, as well as a high cost.

The direct antitumor activity of type I IFN relies on its antiproliferative effect (blocking the G1 phase and lengthening all phases of the cell cycle) and lysis of tumor cells (through apoptosis) [revised in Brassard et al., 2002; Bracarda et al., 2010; Kiladjian et al., 2011]. As immune modulator, type I IFN promotes antitumor action by activating a variety of innate and adaptive immune mechanisms that help kill tumor cells. IFN can increase the expression of surface tumor-associated antigens and other molecules which facilitates its detection by immune surveillance cells [reviewed in Kiladjian et al., 2011], and stimulates the proliferation, activation and generation of CTLs [reviewed in Brassard et al., 2002; reviewed in Bracarda et al., 2010; reviewed in Lasfar et al., 2011] and $CD4^+$ T-cells infiltrating tumors [Hakansson et al., 1996; reviewed in Bracarda et al., 2010]. Treatment with IFN-α activates NK cells, macrophages and dendritic cells (DCs) [reviewed in Lasfar et al., 2011], all of which have antitumor activity.

In addition, IFN may inhibit angiogenesis [Lasfar et al., 2011]. Tumor growth is associated with the formation of new blood vessels, which allows the tumor to obtain energy and nutrients for its growth and metastasis. Inhibition of angiogenesis by IFN-α is related to a decrease in the Basic-Fibroblast Growth Factor (BSGF), a pro-angiogenic factor, and matrix metalloprotease-9 expression, as well as to a reduction of the blood-vessel density in the tumor [Brassard et al., 2002; reviewed in Dec and Puchalski, 2008; reviewed in Bracarda et al., 2010].

In summary, though antiproliferative effects on IFN-α have been described in vitro, it seems that in vivo the modulation of antitumor immune surveillance and inhibition of angiogenesis would be more relevant. However, it is impossible to dissect an isolated biological effect resulting from IFN treatment, as many cell types respond to IFN and also many genes are activated, so it is difficult to accurately determine its in vivo mechanism of antitumor action. In fact, it is possible that its action is a combination of its antiproliferative and antiapoptotic effects, immune modulator action and angiogenesis inhibitor activity [reviewed in Brassard et al., 2002].

In the majority of trials IFN-α is inoculated. One of the main problems of the systemic route in tumor treatment is that the beneficial effects are obtained only at high doses (10-20MIU i.v. repeatedly depending on the protocol), that may exceed the maximum tolerated dose, causing adverse side effects, such as alterations of the nervous system, myelodepression or -suppression [reviewed in Brassard et al., 2002; reviewed in Kiladjian et al., 2011]. Some trials suggest that no results may be observed at these low doses [Brassard et al., 2002], while others discuss that with these low doses its action is more related to its activity on the immune system and angiogenesis than on a direct anti-tumor effect [Bracarda et al., 2010]. As IFN inoculation is expensive and tedious to administer to animals, the low-dose oromucosal route is being explored [Cummins and Pruitt, 1999; reviewed in Dec and Puchalski, 2008]. It has been proposed that oromucosally administered IFN-α can induce beneficial systemic effects in animals, mostly related to the immune system (improvement of $CD4^+$ and $CD8^+$ T-cells counts and of NK cells activity) [Mughini 2000; Cummins et al., 2005; reviewed in Dec and Puchalski, 2008]. In addition, different trials have shown that the oromucosal route is not associated with adverse toxic effects. Although the oral low dose seems very promising as a possible anti-tumor therapy in animals, further studies are needed to confirm its potential for treating different types of tumors such as T-cell or myeloid leukemia, lymphoma, and lymphosarcoma, frequent in feline retrovirosis [De Mari et al., 2004].

Conclusion

Though long-term studies are lacking, it seems that the administration of type I interferon is associated with an improvement of the clinical state of the retrovirus-infected cat. However, cat owners should be aware that this beneficial effect may be transient. This beneficial outcome of interferon therapy is probably due to its effect on the immune system, but viral and proviral loads of infected cats have not proved to be affected by treatment. Moreover, in vitro studies do not sustain an effect of IFN on viral protein synthesis, a fact which is substantiated by the lack of ISREs in the control regions (LTRs) of the feline retrovirus genome. Most studies have been carried out on infected cats with no tumors, and the possible use of interferon for treating the different tumors which may accompany feline retroviral infections is still far from confirmed. To accomplish this, long-term clinical trials should be planified to confirm continued efficacy.

Conflict of Interest Statement

All authors declare no financial or commercial conflict of interest.

Acknowledgements

The authors are indebted to David Bruhn (USA) for his experienced editorial assistance. This work was supported with Spanish AGL2005-05248/GAN, BSCH-UCM GR58/08 (UCM920620), and by Project 20020100100994 of the University of Buenos Aires (Argentina).

References

Agy, MB; Acker, RL; Sherbert, CH; Katze, MG. (1995) Interferon treatment inhibits virus replication in HIV-1- and SIV-infected CD4+ T-cell lines by distinct mechanisms: evidence for decreased stability and aberrant processing of HIV-1 proteins. *Virology*, 214, 379-386.

*American Association of Feline Practitioners (*AAFP) Feline Vaccine Advisory Panel Report. (2006) *JAVMA*, Vol 229, No. 9, Nov 1..

Andrieu, M; Chassin, D; Desoutter, JF; Bouchaert, I; Baillet, M; Hanau, D; Guillet, JG; Hosmalin, A. (2001) Downregulation of major histocompatibility class I on human dendritic cells by HIV Nef impairs antigen presentation to HIV-specific CD8+ T lymphocytes. *AIDS Res. Hum. Retroviruses*, 17(14), 1365-1370.

Arai, M; Earl, DD; Yamamoto, JK. (2002) Is AZT/3TC therapy effective against FIV infection or immunopathogenesis? *Vet. Immunol. Immunopathol.*, 85(3-4), 189-204.

Arjona, A; Escolar, E; Soto, I; Barquero, N; Martin, D; Gomez-Lucia, E. (2000) Seroepidemiological survey of infection by feline leukemia virus and immunodeficiency virus in Madrid and correlation with some clinical aspects. *J. Clin. Microbiol.*, 38(9), 3448-3449.

Baca-Regen, L; Heinzinger, N; Stevenson, M; Gendelman, HE. (1994) A interferon-induced antiretroviral activities: restriction of viral nucleic acid synthesis and progeny virion production in human immunodeficiency virus type 1-infected monocytes. *J. Virol.*, *68*, 7559-7565.

Beatty, JA; Lawrence, CE; Callanan, JJ; Grant, CK; Gault, EA; Neil, JC; Jarrett, O. (1998) Feline immunodeficiency virus (FIV)-associated lymphoma: a potential role for immune dysfunction in tumourigenesis. *Vet. Immunol. Immunopathol.*, 65(2-4), 309-322.

Bednarik, DP; Mosca, JD; Raj, NB; Pitha PM. (1989) Inhibition of human immunodeficiency virus (HIV) replication by HIV-trans-activated alpha 2-interferon. *Proc. Natl. Acad. Sci. USA.*, 86(13), 4958-4962.

Bendinelli, M; Pistello, M; Lombardi, S; Poli, A; Garzelli, C; Matteucci, D; Ceccherini-Nelli, L; Malvaldi, G; Tozzini, F. (1995) Feline immunodeficiency virus: an interesting model for AIDS studies and an important cat pathogen. *Clin. Microbiol. Rev.*, 8(1), 87-112.

Bisset, LR; Lutz, H; Boni, J; Hofmann-Lehmann, R; Luthy, R; Schupbach, J (2002) Combined effect of zidovudine (ZDV), lamivudine (3TC) and abacavir (ABC) antiretroviral therapy in suppressing in vitro FIV replication. *Antiviral Res.*, 53(1), 35-45.

Bracarda, S; Eggermont, AMM; Samuelsson, J. (2010) Redefining the role of interferon in the treatment of malignant diseases. *European J. Cancer*, 46, 284-297.

Brassard, D; Grace, MJ; Bordens, RW (2002) Interferon-α as an immunotherapeutic protein. *J. Leukocyte Biology*, 71, 565-581.

Burkhard, MJ; Dean, GA. (2003) Transmission and immunopathogenesis of FIV in cats as a model for HIV. *Curr. HIV Res.*, 1(1), 15-29.

Callanan, JJ; Jones, BA; Irvine, J; Willett, BJ; McCandlish, IA; Jarrett, O. (1996) Histologic classification and immunophenotype of lymphosarcomas in cats with naturally and experimentally acquired feline immunodeficiency virus infections. *Vet. Pathol.*, 33(3), 264-72.

Caney, S (2000) Feline immunodeficiency virus: an update. *In Practice*, 22 (5), 255-260.

Coccia, EM; Krust, B; Hovanessian, AG. (1994) Specific inhibition of viral protein synthesis in HIV-infected cells in response to interferon treatment. *J. Biol. Chem.*, 269, 23087-23094.

Collado, VM; Doménech, A; Gómez-Lucía, E; Tejerizo, G; Miró, G. (2006) Usos del interferón en la clínica de pequeños animales. *Pequeños Animales*, 63, 68-75.

Collado Alcala, V; Rupérez Moguer, C; Gómez, NV; Miró Corrales, G. (2007a) Tratamiento de las retrovirosis felinas. *Revista Canis et Felis*, 87, 45-54.

Collado, VM; Gómez-Lucía, E; Tejerizo, G; Miró, G; Escolar, E; Martín, S; Doménech, A. (2007b) Effect of type I interferons on the expression of feline leukaemia virus. *Vet. Microbiol.*, 123, 180–186.

Cotter, SM. (1979). Anaemia associated with feline leukemia virus infection. *J. Am. Vet. Med. Assoc.*, 175(11), 1191-1194.

Cummins, MJ; Pruitt, B. (1999) Low-dose oral use of human interferon-alpha in cancer patients. *J. Interferon Cytokine Res.*, 19, 937-941.

Cummins, JM; Tompkins, MB; Olsen, RG; Tompkins, WA; Lewis, MG. (1988) Oral use of human alpha interferon in cats. *J. Biol. Response Mod.*, 7(5), 513-523.

Cummins, JD; Krakowka, GS; Thompson, CG. (2005) Systemic effects of interferons after oral administration in animals and humans. *Am. J. Vet. Res.*, 66, 164–176.

De Mari, K; Maynard, L; Sanquer, A; Lebreux, B; Eun, HM. (2004) Therapeutic effects of recombinant feline interferon-omega on feline leukemia virus (FeLV)- infected and FeLV/feline immunodeficiency virus (FIV)-coinfected symptomatic cats. *J. Vet. Intern. Med.*, 18, 477-482.

De Rozières, S; Swan, CH; Sheeter, DA; Clingerman, KJ; Lin, YC; Huitron-Resendiz, S; Henriksen, S; Torbett, BE, Elder, JH. (2004) Assessment of FIV-C infection of cats as a function of treatment with the protease inhibitor, TL-3. *Retrovirology*, 1, 38-50. doi:10.1186/1742-4690-1-38.

Dec, M; Puchalski, A. (2008) Use of oromucosally administered IFN-α in the prevention and treatment of animal diseases. *Polish Journal of Veterinary Sciences*, 11, 171-182.

Dianzani, F; Castilletti, C; Gentile, M; Gelderblom, HR; Frezza, F; Capobianchi, MR. (1998) Effects of IFN α on late stages of HIV-1 replication cycle. *Biochimie*, 80, 745-754.

DiCenzo, R; Peterson, D; Cruttenden, K; Morse, G; Riggs, G; Gelbard, H; Schifitto, G. (2004) Effects of valproic acid coadministration on plasma efavirenz and lopinavir concentrations in human immunodeficiency virus-infected adults. *Antimicrob. Agents Chemother.*, 48(11), 4328–4331.

Dietrich, I; Hosie, MJ; Willett, BJ. (2011) The role of BST2/tetherin in feline retrovirus infection. *Vet. Immunol. Immunopathol.*, 143(3-4), 255-64.

Doménech, A; Collado, VM; Martín, S; Gómez-Lucía E. (2007) Patogenia de la leucemia y la inmunodeficiencia felinas. *Revista Canis et Felis*, 87, 36-54.

Doménech, A; Tejerizo, G; Illera, JC; Gómez-Lucía, E. (2009) Effect of male steroid hormones on retroviral infections. *Current Topics in Steroid Research* 6, 69-81.

Doménech, A; Miró, G; Collado, VM; Ballesteros, N; Sanjosé, L; Escolar, E; Martin, S; Gomez-Lucia, E. (2011) Use of recombinant interferon omega in feline retrovirosis: from theory to practice. *Vet. Immunol. Immunopathol.*, 143(3-4), 301-306. doi:10.1016/j.vetimm. 2011.06.008.

Dunham, SP; Graham, E. (2008) Retroviral infections of small animals. *Vet. Clin. Small Anim.*, 38, 879-901.

Feng, X; Heyden, NV; Ratner, L. (2003) Alpha interferon inhibits human T-cell leukemia virus type 1 assembly by preventing Gag interaction with rafts. *J. Virol.*, 77, 13389–13395.

Feldman, S; Stein, D; Amrute, S; Denny, T; Garcia, Z; Kloser, P; Sun, Y; Megjugorac, N; Fitzgerald-Bocarsly P. (2001) Decreased interferon-alpha production in HIV-infected patients correlates with numerical and functional deficiencies in circulating type 2 dendritic cell precursors. *Clin. Immunol.*, 101(2), 201-210.

Fernandez, S; Tanaskovic, S; Helbig, K; Rajasuriar, R; Kramski, M; Murray, JM; Beard, M; Purcell, D; Lewin, SR; Price, P; French, MA *(2011)* CD4+ T-cell deficiency in HIV patients responding to antiretroviral therapy is associated with increased expression of interferon-stimulated genes in CD4+ T cells. *J. Infect. Dis.*, 204(12), 1927-1935.

Finke, JS; Shodell, M; Shah, K; Siegal, FP; Steinman, RM. (2004) Dendritic cell numbers in the blood of HIV-1 infected patients before and after changes in antiretroviral therapy. *J. Clin. Immunol.*, 24(6), 647-652.

Fletcher, N; Meeker, RB; Hudson, LC; Callanan, JJ. (2011) The neuropathogenesis of feline immunodeficiency virus infection: Barriers to overcome. *Vet. J.*, 188(3), 260–269.

Flynn, JN; Cannon, CA; Lawrence, CE; Jarret, O. (1994) Polyclonal B-cell activation in cats infected with feline immunodeficiency virus. A. *Immunology*, 81, 626-630.

Gabor, LJ; Love, DN; Malik, R; Canfield, PJ. (2001). Feline immunodeficiency virus status of Australian cats with lymphosarcoma. *Aust. Vet. J.*, 79, 540-545.

Gendelman, HE; Baca, LM; Turpin, J; Kalter, DC; Hansen, B; Orenstein, JM; Dieffenbach, W; Friedman, RM; Meltzer, MS. (1990) Regulation of HIV replication in infected monocytes by IFN-alpha. Mechanisms for viral restriction. *J. Immunol.*, 145(8), 2669-2676.

Gisbert, MA; Gómez, N. (2010) Comparison of different therapies in cats spontaneously infected with Feline Immunodeficiency Virus (FIV): Zidovudine versus Zidovudine associated to Valproic Acid. *International Feline Retrovirus Research Symposium (IFRRS)* Charleston, SC, USA, May 23–26.

Gleich, S; Hartmann, K.(2009) Hematology and serum biochemistry of feline immunodeficiency virus-infected and feline leukemia virus-infected cats. *J. Vet. Intern. Med.*, 23(3), 552-558.

Gomez, NV (2000) Feline immunodeficiency virus (FIV): advances in the diagnosis and treatment of feline AIDS. *Revista Veterinaria Argentina*, 17 (166), 468-477.

Gomez, NV; Gisbert, MA; Jáuregui, L. (2002) Efecto terapéutico de la zidovudina (AZT) en pacientes con retrovirosis felinas. *Revista Veterinaria Argentina*, 19 (183), 224-236.

Gomez-Lucia, E; Collado, VM; Miró, G; Doménech, A.(2009) Effect of type-I interferon on retroviruses. *Viruses*, 1(3), 545-573.

Håkansson, A; Gustafsson, B; Krysander, L; Håkansson, L. (1996) Tumour-infiltrating lymphocytes in metastatic malignant melanoma and response to interferon alpha treatment. *Br. J. Cancer*, 74, 670–676.

Haller, O; Kochs, G; Weber, F. (2006) The interferon response circuit: induction and suppression by pathogenic viruses. *Virology*, 344, 119-130.

Hansen, BD; Nara, PL; Maheshwari, RK; Sidhu, GS; Bernbaum, JG; Hoekzema, D; Meltzer, MS; Gendelman, HE. (1992) Loss of infectivity by progeny virus from alpha interferon-treated human immunodeficiency virus type 1-infected T cells is associated with defective assembly of envelope gp120. *J. Virol.*, 66(12), 7543-7548.

Hartmann, K. (2006) Feline leukaemia virus infection. In: Greene, C.E. (Ed.), Infectious Diseases of the dog and cat, 3ª ed., Saunders Elsevier, St. Louis, pp 105-131.

Hartmann, K. (2011) Clinical aspects of feline immunodeficiency and feline leukemia virus infection. *Vet. Immunol. Immunopathol.*, 143(3-4), 190-201. doi:10.1016/ j.vetimm.2011.06.003.

Hofmann-Lehmann, R; Cattori, V; Tandon, R; Boretti, FS; Meli, ML; Riond, B; Pepin, AC; Willi, B; Ossent, P; Lutz H. (2007) Vaccination against the feline leukaemia virus: outcome and response categories and long-term follow-up. *Vaccine*, 25, 5531-5539.

Honda, K; Yanai, H; Takaoka, A; Taniguchi, T. (2005) Regulation of the type I IFN induction: a current view. *Int. Immunol.*, 17, 1367-1378.

Hosie, MJ; Addie, D; Belák, S; Boucraut-Baralon, C; Egberink, H; Frymus, T; Gruffydd-Jones, T; Hartmann, K; Lloret, A; Lutz, H; Marsilio, F; Pennisi, MG; Radford, AD; Thiry, E; Truyen, U; Horzinek, MC. (2009) Feline immunodeficiency. ABCD guidelines on prevention and management. *J. Feline Med. Surg.*, 11, 575-584.

Hosmalin, A; Lebon P. (2006) Type I interferon production in HIV-infected patients. *J. Leuk. Biol.*, 80, 984-993.

Kiladjian, JJ; Mesa, RA; Hoffman R. (2011) The renaissance of interferon therapy for the treatment of myeloid malignancies. *Blood*, 117, 4706-4715.

Lasfar, A; Abushahba, W; Balan, M; Cohen-Solal KA. (2011) Interferon lambda: a new sword in cancer immunotherapy. *Clin. Dev. Immunol.* doi:10.1155/2011/349575.

Lawson, MH; Hosie, MJ. (2001) Follow-up of feline immunodeficiency virus-infected cats. *Vet. Rec.*, 148 (14), 449-450.

Lehman, TL; O'Halloran, KP; Fallon, SA; Habermann, LM; Campbell, JA; Nordone, S; Dean, GA; Hoover, EA; Avery, PR. (2009) Altered bone marrow dendritic cell cytokine production to toll-like receptor and CD40 ligation during chronic feline immunodeficiency virus infection. *Immunology*, 126, 405-412.

Levy, JK; Ritchey, JW; Rottman, JB; Davidson, MG; Liang, Y; Jordan, HL; Tompkins, WA; Tompkins, MB. (1998) Elevated interleukin-10-to-interleukin-12 ratio in feline immunodeficiency virus-infected cats predicts loss of type 1 immunity to Toxoplasma gondii. *J. Infect. Dis.*, 178, 503-511.

Levy, JK; Liang, Y; Ritchey, JW; Davidson, MG; Tompkins, WA; Tompkins, MB. (2004) Failure of FIV-infected cats to control *Toxoplasma gondii* correlates with reduced IL2, IL6, and IL12 and elevated IL10 expression by lymph node T cells. *Vet. Immunol. Immunopathol.*, 98,101-11.

Little, S; Sears, W; Lachtara, J; Bienzle, D. (2009) Seroprevalence of feline leukemia virus and feline immunodeficiency virus infection among cats in Canada. *Can. Vet. J.*, 50:644-648.

Lutz, H; Addie, D; Belak, S; Boucraut-Baralon, C; Egberink, H; Frymus, T; Gruffydd-Jones, T; Hartmann, K; Hosie, MJ; Lloret, A; Marsilio, F; Pennisi, MG; Radford, AD; Thiry, E; Truyen, U; Horzinek, MC. (2009) Feline leukaemia. ABCD guidelines on prevention and management. *J. Feline Med. Surg.*, 11, 565-574.

Marié, I; Durbin, JE; Levy, DE. (1998) Differential viral induction of distinct interferon-α genes by positive feedback through interferon regulatory factor-7. *EMBO J.*, 17, 6660-6669.

McCaw, DL; Boon, GD; Jergens, AE; Kern, MR; Bowles, MH; Johnson, JC. (2001) Immunomodulation therapy for feline leukemia virus infection. *J. Am. Anim. Hosp. Assoc.*, 37, 356–363.

Meylan, PR; Guatelli, JC; Munis, JR; Richman, DD; Kornbluth, RS. (1993) Mechanisms for the inhibition of HIV replication by interferons-α, -beta, and -gamma in primary human macrophages. *Virology*, 193, 138-148.

Mogensen, KE; Lewerenz, M; Reboul, J; Lutfalla, G; Uzé, G. (1999) The type I interferon receptor: structure, function, and evolution of a family business. *J. Interferon. Cytokine Res.*, 19, 1069-1098.

Mughini, L. (2000) Effects of interferon alpha administered by peroral route on lymphomonocytes' subpopulations of peripheral blood in healthy volunteers. *Clin. Ter.*, 151(1 Suppl 1), 3-12.

Pedretti, E; Passeri, B; Amadori, M; Isola, P; Di Pede, P; Telera, A; Vescovini, R; Quintavalla, F; Pistello, M. (2006) Low-dose interferon-alpha treatment for feline immunodeficiency virus infection. *Vet. Immunol. Immunopathol.*, 109(3-4), 245-254.

Phillips, K; Arai, M; Tanabe, T; Raskin, R; Volz, M; Uhl, EW; Yamamoto, JK. (2005) FIV-infected cats respond to short-term rHuG-CSF treatment which results in anti-G-CSF neutralizing antibody production that inactivates drug activity. *Vet. Immunol. Immunopathol.*, 108(3-4), 357-371.

Poli, G; Orenstein, JM; Kinter, A; Folks, TM; Fauci AS. (1989) Interferon-α but not AZT suppresses HIV expression in chronically infected cell lines. *Science*, 244, 575-577.

Poss, M; Ross, H. (2008) Evolution of the long terminal repeat and accessory genes of feline immunodeficiency virus genomes from naturally infected cougars. *Virology* 370, 55-62.

Power, C; Zhang, K.; van Marle, G. (2004) Comparative neurovirulence in lentiviral infections: the roles of viral molecular diversity and select proteases. *J. Neurovirol.*, 10, 113–117.

Riondato, F; Gianella, P; Guglielmino, R; Cagnasso, A; Bo, S. (2003) Effects of interferon alpha (INF-alpha) therapy on peripheral blood lymphocyte subsets from FIV and FeLV naturally infected cats. *Vet. Res. Commun.*, 27 Suppl 1, 429-432.

Robert-Tissot, C; Rüegger, VL; Cattori, V; Meli, ML; Riond, B; Gomes-Keller, MA; Vögtlin, A; Wittig, B; Juhls, C; Hofmann-Lehmann, R; Lutz, H. (2011) The innate antiviral immune system of the cat: molecular tools for the measurement of its state of activation. *Vet. Immunol. Immunopathol.*, 143(3-4):269-81.

Samarijiwa S., personal communication. Dr. Samarajiwa, one of the developers of the software Interferome, described in Samarajiwa, SA; Forster, S; Auchettl, K; Hertzog, PJ. (2009) Interferome: the database of interferon regulated genes. *Nucleic Acids Res.*, 37(Database issue), D852-857.

Savarino, A; Pistello, M; D'Ostilio, D; Zabogli, E; Taglia, F; Mancini, F; Ferro, S; Matteucci, D; De Luca, L; Barreca, ML; Ciervo, A; Chimirri, A; Ciccozzi, M; Bendinelli, M. (2007) Human immunodeficiency virus integrase inhibitors efficiently suppress feline immunodeficiency virus replication in vitro and provide a rationale to redesign antiretroviral treatment for feline AIDS. *Retrovirology*, 30(4), 79-91. doi:10.1186/1742-4690-4-79.

Sellon, R; Hartmann, K. (2006) Feline Immunodeficiency Virus Infection. *In*: Greene, C.E. (Ed.), Infectious Diseases of the Dog and Cat, WB Saunders, St. Louis, pp. 131 - 142.

Shimode, S; Miyazawa, T; Kobayashi, T; Sato, H; Tanabe, T. (2012) Identification and Characterization of Feline UBE1L Gene. *J. Vet. Med. Sci.* 2012, 74(2), 235-239.

Shirazi, Y; Pitha, PM. (1992) Alpha interferon inhibits early stages of the human immunodeficiency virus type 1 replication cycle. *J. Virol.*, 66, 1321-1328.

Smith, MS; Thresher, RJ; Pagano, JS. (1991) Inhibition of human immunodeficiency virus type 1 morphogenesis in T cells by α interferon. *Antimicrob. Agents Chemother.*, 35, 62-67.

Soumelis, V; Scott, I; Gheyas, F; Bouhour, D; Cozon, G; Cotte, L; Huang, L; Levy, JA; Liu, YJ. (2001) Depletion of circulating natural type 1 interferon-producing cells in HIV-infected AIDS patients. *Blood*, 98(4), 906-912.

Steed, VP. (1987) Improved survival of four cats infected with feline leukemia virus after oral administration of interferon. *Feline Pract.*, 17, 24–30.

Tejerizo, G; Domenech, A; Illera, JC; Collado, VM; Gomez-Lucia, E. (2005) Effect of 17beta-estradiol and progesterone on the expression of FeLV in chronically infected cells. *Vet. Microbiol.*, 109(3-4), 191-199.

Torbett, BE; Elder, JH; Swan, CH; de Rozières, S; Clingerman, KJ; Hoover, E. (2003) The effects of the broad-based protease inhibitor TL-3 in FIV-C infected cats. *Current Drug Targets Infect. Dis.*, (3), 295-309.

Uckun, FM; Chen, CL; Samuel, P; Pendergrass, S; Venkatachalam, TK; Waurzyniak, B; Qazi, S. (2003) In vivo antiretroviral activity of stampidine in chronically feline immunodeficiency virus-infected cats. *Antimicrob. Agents Chemother.*, 47(4), 1233–1240.

Weiss, RC; Cummins, JM; Richards, AB. (1991) Low-dose orally administered alpha interferon treatment for feline leukemia virus infection. *J. Am. Vet. Med. Assoc.*, 199, 1477-1481.

Willett, BJ; Jarrett, O. (1995) Feline immunology and immunodeficiency. Oxford Science Publications. Glasgow.

Williams, BR. (1991) Signal transduction and transcriptional regulation of interferon-alpha-stimulated genes. *J. Interferon Res.*, 11, 207-213.

Xue Q, Yang L, Liu X, Liu W. (2010) Molecular characterization of feline type I interferon receptor 2. *J. Interferon. Cytokine Res.*, 30(2): 81-8.

Zeidner, NS; Myles, MH; Mathiason-DuBard, CK; Dreitz, MJ; Mullins, JI; Hoover, EA. (1990) Alpha interferon (2b) in combination with zidovudine for the treatment of presymptomatic feline leukemia virus-induced immunodeficiency syndrome. *Antimicrob. Agents Chemother.*, 34(9), 1749-1756.

In: Felines
Editors: R. García and R. Jiménez
ISBN 978-1-62257-005-8

Chapter 2

EMERGING TICK-BORNE PROTOZOAL DISEASES - PART 1: NEW ADVANCES IN DIAGNOSIS, EPIZOOTIOLOGY AND TAXONOMY OF FELINE PIROPLASMIDS

Angel Criado-Fornelio
Parasitology Laboratory. Microbiology and Parasitology Department, Faculty of Pharmacy, Universidad de Alcalá, Alcalá de Henares, Spain

ABSTRACT

In vertebrates, piroplasmid protozoa infect blood cells. Both wild felines and domestic cats may be infected by *Babesia* and *Cytauxzoon* species. Feline piroplasmosis is assumed to be tickborne, but the definitive host has not been identified in many piroplasmid species. Babesiosis and cytauxzoonosis are considered as emerging diseases, affecting felids in a wide geographic range. Usually these infections are asymptomatic but in some cases disease is evident and animals may eventually die. This is much more likely whenever concurrent pathogens (such as *Mycoplasma* or viruses) attack felids. It must be emphasized that babesiosis is present in wild felines, included those qualified as endangered species. Piroplasmids may be a severe threat for the Iberian lynx (*Lynx pardina*) and the Serengeti lion (*Panthera leo*). Although piroplamosis may be treated with primaquine, imidocarb or some other drugs, chemotherapy results are not always satisfactory.

Before the advent of molecular diagnosis methods, piroplasmid species infecting felines were limited to *Babesia cati*, *B. felis*, *B. herpailuri*, *B. pantherae* and *Cytauxzoon felis*. However, some contributions published in the last years have shown that there is a broader spectrum, which includes *Theileria annae* (or *Babesia microti*-like), *B. leo*, *B. canis canis*, *B.canis presentii*, *C. manul*, *Babesia* sp, *B. canis vogeli* and *B. lengau*. The classification of feline piroplasmids is in continuous evolution due to the application of molecular techniques such as DNA sequencing. In agreement with the opinion expressed by some authors, and based on current biological and epizootiological data, it is evident that the species *Theileria annae* (= *Babesia microti*-like) should be denominated *Babesia annae*.

Microscopic examination of blood smears has proven to be a low sensitivity method for detecting piroplasmid infection. New diagnostic procedures like the Polymerase Chain Reaction (PCR) and the Reverse Line Blotting assay (RLB) have been implemented to improve diagnosis. Up to date, no real-time PCR assays have been published for diagnosis of piroplasmosis in cats. By this reason, in this chapter is described a new method for diagnosis of feline piroplasmids and *Hepatozoon* sp. The procedure is based on quantitative PCR with FRET probes.

1. Introduction

Feline tick-borne protozoal diseases are mainly caused by piroplasmida and *Hepatozoon* sp. In this chapter, we present a review of known piroplasmid pathogens of wild and domestic felines. Information related to hepatozoons is reserved to the next chapter (Emerging tick-borne protozoal diseases - part 2).

Piroplasmid protozoa have a complex life-cycle which includes two hosts: the definitive host (Ixodid ticks) and the intermediate host (diverse vertebrates - including cats and their kin). In felines, parasites live in blood cells, causing sporadic cases of disease, and more rarely outbreaks (Carli *et al.*, 2012).

First babesial infections were described in ruminants of Romania at the end of the XIX century (Solano-Gallego and Baneth, 2011). However, babesid infections in felines were discovered much later (Davis, 1929). Many authors consider these tick-borne diseases as emerging zoonoses (Meinkoth and Kocan, 2005; Holman and Snowden, 2010).

The frequency of babesiosis is lower in felines than in canids. However, sometimes feline piroplasmosis may be life-threatening in domestic cats and wild felids (Munson *et al.*, 2008). Disease may develop after long incubation period making difficult an accurate diagnosis. In addition, concomitant pathogen infection (caused by *Mycoplasma* spp or viruses) usually complicate prognosis. To worsen the situation, the clinical signs of piroplasmosis (*e.g.* fever, lethargy, anorexia and anaemia) are not always unequivocal (Schoeman *et al.*, 2001; Carli *et al.*, 2012). Such disease expression may be confounded with *Mycoplasma* infection signs. Luckily, it seems that feline piroplasmids are not a concern for humans, although some authors have pointed that there is a potential zoonotic danger for some pathogens like *Babesia (Theileria) annae* (Camacho *et al.*, 2001 and 2002).

About one dozen of piroplasmid species (corresponding to the genera *Babesia* and *Cytauxzoon*) have been described in felines based on morphological or/and molecular grounds (Penzhorn, 2006, Solano-Gallego and Baneth, 2011). In general their geographic distribution and host range (both definitive and intermediate) is largely unknown (Criado-Fornelio *et al.*, 2003a and b). Diagnosis of feline piroplasmosis is facilitated by the use of molecular methods (Criado-Fornelio *et al.*, 2004). Current molecular detection procedures are standard PCR (Penzhorn *et al.*, 2001), seminested PCR (Criado-Fornelio *et al.*, 2003a), Reverse Line Blotting - RLB (Bosman *et al.*, 2007; Georges *et al.*, 2008) and nested PCR (Simking *et al.*, 2010).

In the present chapter, we shall review different features of feline piroplasmids, such as the morphology, biology, pathology, epidemiology, diagnosis, classification and control measures to reduce the risk of disease transmission.

2. Feline Piroplasmids: Morphology and Life Cycle

Feline *Babesia* and *Cytauxzoon* isolates have been morphologically classified as small (length < 1.5 μm) or large (length > 1.5 μm) species based on their aspect when the parasites are observed inside host erythrocytes (the same is true for non-feline piroplasmids). Most feline babesids do not form maltese cross inside red blood cells. An overview of piroplasmid species found in wild and domestic felines is shown in Table 1. In the past, the lack of molecular studies has created a certain degree of confusion in feline piroplasm taxonomy (Penzhorn, 2006). As an example, *Babesiella felis* (Carpano, 1934) is considered a synonym of *B. felis* (Penzhorn *et al.*, 2001).

Table 1. Main features of known feline piroplasmid species

Species	Size / Maltese cross	Host(s) / Diagnostic method	Publication - country
Babesia cati	Large / No	*Felis catus* - Wild cat / Microscopy	Mudaliar *et al.*, (1950) - India
B. herpailuri	Large / No	*Herpailurus jaguarundi* / Microscopy	Dennig, (1967) – Venezuela
B. pantherae	Large / No	*Panthera pardus* / Microscopy	Dennig *et al.*, (1972) - Kenya
B. canis vogeli	Parasite not observed by microscopy	Domestic cat / RLB + DNA sequencing	Georges *et al.*, (2008) - Trinidad
B. canis vogeli?	Large / No	Domestic cat / Microscopy + DNA sequencing	Simking *et al.*, (2010) - Thailand
B. felis = *Babesiella felis?* (Carpano, 1934)	Small / Yes	*Felis sylvestris* / Microscopy (Davis, 1929) and sequencing (Penzhorn *et al.*, 2001).	Davis, (1929) – Sudan and Penzhorn *et al.*, (2001) – South Africa
B. leo	Small / Yes	*Panthera leo* / Microscopy + sequencing	Penzhorn *et al.*, (2001) – South Africa
Cytauxzoon felis	Small / No	Domestic cat / Microscopy	Wagner, (1976) - USA
Cytauxzoon sp	Small / No	Domestic cat / DNA sequencing (Criado-Fornelio *et al.*, (2004) Microscopy + DNA sequencing(Millán *et al.*, (2007)	Criado-Fornelio *et al.*, (2004) and Millán *et al.*, (2007) - Spain
C. manul	Small / No	*Otocolobus manul* / Microscopy + sequencing	Reichard *et al.*, (2005) - Mongolia
Babesia sp	Small / No	*Puma concolor* / Microscopy + sequencing	Yabsley *et al.*, (2006) - USA
B. canis presentii	Small / No	Domestic cat / Microscopy + sequencing	Baneth *et al.*, (2004) – Israel
B. canis canis	Parasite not observed by microscopy	Domestic cat / Sequencing	Criado-Fornelio *et al.*, (2003a) - Spain
Babesia (*Theileria*) *annae* (*B. microti*-like)	Parasite not observed by microscopy	Domestic cat / Sequencing	Criado-Fornelio *et al.*, (2003a) - Portugal
B. lengau	Small / No	*Acinonyx jubatus* / Microscopy + sequencing	Bosman *et al.*, (2010) – South Africa

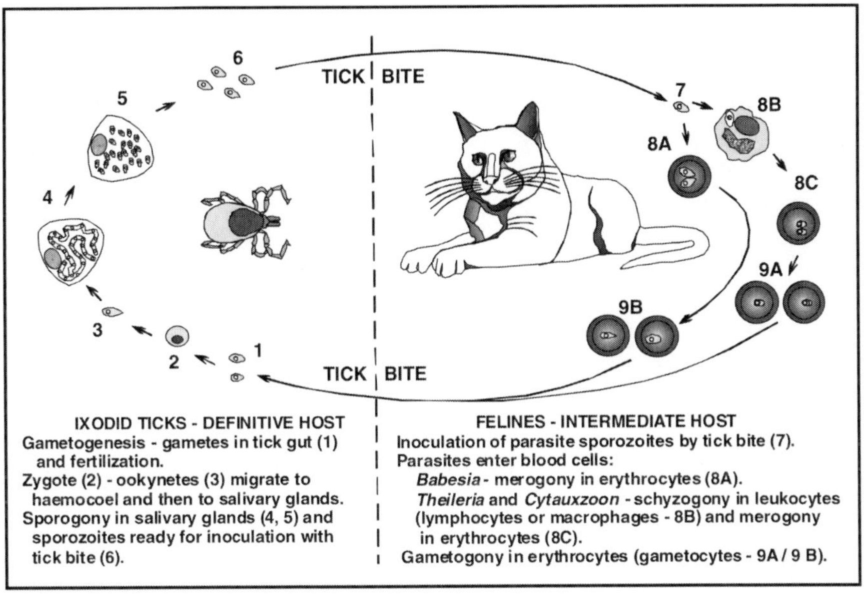

Figure 1. Schematic representation of the life cycle of piroplasmids (*Babesia*, *Theileria*, *Cytauxzoon*).

The life cycle of feline piroplasmids (Figure 1) involves two types of host. Cats or other felines serve as intermediate hosts while Ixodid ticks are definitive hosts (Schmidt and Roberts, 1981). During a blood meal, a *Babesia*-infected tick introduces sporozoites into the feline host. Inoculated protozoa enter in erythrocytes (with the aid of a specialized apical complex, an outstanding morphological feature of the Apicomplexa) and undergo asexual reproduction. In the case of *Cytauxzoon* sp, parasites have a pre-erythrocytic stage in lymphocytes, although some studies indicate that the cell type most frequently infected is the macrophage (Kier, 1987; Kocan *et al.*, 1992; Susta *et al.*, 2009). There is a *Theileria* species capable of infecting cats (*T. annae* as reported by Criado-Fornelio *et al.*, 2003a). However, there are some doubts in the adscription of this species to the genus *Theileria*, since the protozoa has not been found in dog lymphocytes or macrophages (Camacho *et al.*, 2001, Goethert and Telford, 2003). Going back to the life-cycle of piroplasmids, the parasites differentiate into gametocytes after some cycles of merogony in vertebrate blood cells. Once gametocytes are ingested by the tick, protozoa gametes develop in the arthropod gut and then fuse to form a zygote. The diploid cell undergoes a multiple fission process in tick gut (sporogony). The sporozoites migrate (through the coelom) to the tick salivary glands where they are able to restart the life cycle, once a new bite occurs (Kumar *et al.*, 2008). Both trans-stadial and trans-ovarian transmission has been found in *Babesia* species. In contrast, trans-ovarian transmission has not been reported for any *Theileria* or *Cytauxzoon* species. It seems that all those biological features allow the parasite to remain inside ticks for a high number of generations. Hence, parasite survival is strongly favoured (Solano-Gallego and Baneth, 2011).

3. Feline Piroplasmids: Pathology

The presentation of feline piroplasmosis is not always consistent. Table 2 shows the main clinical findings reported in symptomatic cases of feline babesiosis. It is important to underline that much of this information is mainly based in studies on *Cytauxzoon* sp and *Babesia felis*.

Some of the most common signs of feline piroplasmosis (with their score) are the following:

- Anaemia (7)
- Icterus (jaundice) (6)
- Pyrexia (5)
- Anorexia (5)
- Lethargy/depression (5)
- Haematuria (2)
- Dyspnea (1)

The score (number in parentheses) is the frequency of these clinical signs according to data in Table 2. In this ranking, pyrexia is relatively common, which is in contrast with the findings of Futter and Belonje (1980a) and Schoeman *et al.*, (2001). Such difference may have been caused by the fact that these authors studied cats infected by *B. felis*, whilst other researchers have studied mainly felines infected by *Cytauxzoon* or *Babesia* species.

Some of the haematological parameters frequently altered by *B. felis* infection are erythrocyte count and hematocrit (Kocan *et al.*, 1985; Futter and Belonje, 1980b; Rotstein *et al.*, 1999; Schoeman *et al.*, 2001). In some instances thrombocytopenia and changes in leukocyte counts have been reported as well (Schoeman *et al.*, 2001; Baneth *et al.*, 2004), but these hematological changes do not always occur.

Table 2. Summary of clinical findings reported in cases of feline piroplasmosis

Publication - country	Host/ Piroplasmid	Clinical signs	Concurrent infections
Mudaliar *et al.*, (1950) – India	Cat / *Babesia* sp	Pyrexia	None
Stewart *et al.*, (1980) – South Africa	Domestic cat / *Babesia* sp	Anaemia, low haemoglobin and haematocrit, neutrophilia.	None
Hoover et al., (1994) - The USA	Domestic cat / *Cytauxzoon*	Anemia, lethargy, anorexia, icterus, depression, dehydration. Increased ALT, bilirubin and glucose.	Not studied
Schoeman *et al.*, (2001) – South Africa	Domestic cat / *B. felis*	Depression, anaemia, thrombocytopaenia, increased ALT, ALP, urea, creatinin and bilirubin. Icterus and anorexia were less common signs	> 10% coinfected with either *Mycoplasma* sp, FeLV or FIV.
Baneth *et al.*, (2004) - Israel	Domestic cat / *B. canis presentii*	Pyrexia, icterus, lethargy, anaemia, haematuria, thrombocytopaenia, leukopenia, increased ALT, LDH and creatin kinase	FIV
Birkenheuer *et al.*, (2006b) – The USA	Domestic cat / *Cytauxzoon felis*	Fever, icterus, pancytopenia.	Not studied
Weisman *et al.*, (2007) - The USA	Domestic cat / *C. felis*	Abortion?	None
Harvey *et al.*, (2007) – The USA	Cougar / *C. felis*	Low haematocrit, anaemia, increased bilirubin, ALT and AST levels	None
Snider *et al.*, (2010)	Domestic cat / *C. felis*	Dyspnea, fever, icterus, anorexia and lethargy	Not studied
Baraboglia (2011) – Argentina	Domestic cat / *C. felis*	Pyrexia, anorexia, depression, anaemia, hematuria	*Mycoplasma* sp
Carli *et al.*, (2012) – Italy	Domestic cat / *Cytaux-zoon* sp	Pyrexia, anaemia, diarrhoea, vomiting, depression, anorexia and weight loss, jaundice, nervous signs. Increased GGT, bilirubin, urea and glucose.	FIV

Diverse histopathological studies in cats have revealed that some piroplasmid species like *C. felis* produces significant morphologic changes in many host organs and tissues (Susta *et al.*, 2009; Snider *et al.*, 2010). Moreover, in piroplasmid-infected felines, several blood biochemistry parameters are altered, which is a further evidence of tissue damages (Kocan *et al.*, 1985). These changes include increased LDH (Lactate Dehydrogenase), GGT (Gamma-Glutamyl transpeptidase), ALT (Alanine Aminotransferase) or ALP (Alkaline Phosphatase). All these anomalies in blood chemistry are indicative of damage in liver or other tissues. In addition, increased levels of creatinin and urea may occur (which is indicative of kidney malfunction). The hemoglobin content is usually reduced. In one case, abortion and death occurred in an infected female cat (Weisman *et al.*, 2007), although it was not evident the existence of a fetal infection. However, abortion linked to fetal invasion by *Babesia* parasites has been reported in other mammals (such as wild boar, according to Zobba *et al.*, 2011). The survival rate of cats infected by piroplasmids may be influenced by protozoa strains (virulent or avirulent, according to Brown *et al.*, 2009).

Feline piroplasmosis prognosis is complicated by the occurrence of concomitant infections. Pathogens such as *Mycoplasma haemofelis* and *M. haemominutum* have been observed in cats infected by piroplasmids (Baneth *et al.*, 1998; Schoemann *et al.*, 2001; Criado-Fornelio *et al.*, 2003a; Kumar *et al.*, 2008; Baraboglia, 2011). The same is true for infections casued by the Feline Immunodeficiency Virus (FIV) and the Feline Leukemia Virus (FeLV) (Baneth *et al.*, 1998; Schoemann *et al.*, 2001; Criado-Fornelio *et al.*, 2003a; Baneth *et al.*, 2004, Kumar *et al.*, 2008 and Carli *et al.*, 2012).

4. Feline Piroplasmosis: Advances in Diagnosis

4.1. Microscopy

The traditional diagnosis method for feline piroplasmids is microscopy. As stated above, *Cytauxzoon* and *Babesia* merozoites may be observed in feline blood, but unfortunately, parasitemia is usually low (Carli *et al.*, 2012). Hence, microscopic observation usually fails to detect the parasite (Criado-Fornelio *et al.*, 2003a,b). Some reports indicate that detection of piroplasmids in dogs is improved when blood is obtained from capillaries (Bohm *et al.*,

2006), and the same is probably true for felines (Schoeman *et al.*, 2001). Although the use of the buffy coat (or quantitative buffy coat) method is recommended for detection of blood pathogens in humans (Gray *et al.*, 1991), these techniques have not been used in felines for routine diagnosis (Solano and Baneth, 2011), probably due to economic reasons. As stated under the "Morphology and Biology" section in this chapter, microscopic examination of stained blood smears is not an efficient procedure in order to determine the infecting piroplasmid species. Hence, the employ of molecular diagnostic methods is preferred (Penzhorn, 2006).

4.2. Serology

Immunological methods have been not extensively applied to detection of piroplasmid infection in felines. Some tests based on IFAT (Immuno Fluorescent Antibody Technique) have been developed for the detection of antibodies against *C. felis* (Shindel, 1978; Cowell *et al.*, 1988; Filoni *et al.*, 2012) or *B. felis* (López-Rebollar *et al.*, 1999). However, these tests are mainly experimental and are not available commercially.

4.3. DNA-based Methods

The 18S rRNA gene is the preferred target for molecular diagnosis in feline piroplasmids (and likewise for many other organisms). Allsopp *et al.*, (1994) were the first authors to sequence the 18S rRNA gene of a feline piroplasmid (*C. felis*). Meinkoth *et al.*, (2000) studied *C. felis* infection in cats from the USA. Other feline piroplasmids (*B. leo* and *B. felis*) were studied by Penzhorn *et al.*, (2001). However, *B. leo* parasites had been previously detected by microscopy (Lopez-Rebollar *et al.*, 1999). In a molecular survey performed in domestic cats by Criado-Fornelio *et al.*, (2003a), the presence of *Babesia canis canis* and *Babesia (Theileria) annae* was reported in cats from Portugal. *B. canis presentii*, a new subspecies closely related to *B. canis canis,* was diagnosed by molecular methods by Baneth *et al.*, (2004) in Israel.

Diverse PCR methods have been subsequently applied for detection of feline piroplasmids. Birkenheuer *et al.*, (2006a) published a new PCR assay designed for the detection of *Cytauxzoon felis* only. The method is highly specific, so that other feline piroplasmids would remain undetected by employing such a procedure. Yabsley *et al.*, (2006) and Tabar *et al.*, (2008) have designed primers for diagnosis of piroplasmosis in Florida cougars and domestic cats, respectively. The couple of primers PIRO-F / PIRO-R (primarily intended for detection of piroplasmids in dogs by Ano *et al.*, 2001), has been also used with success in cats (Simking *et al.*, 2010). The primers designed by Meinkoth et al (2000) have been employed also by other researchers like Reichard *et al.*, 2005 and Millán *et al.*, (2007 and 2009) for diagnosis of *Cytauxzoon* sp. There had been at least one attempt to develop a multiplex PCR assay for pathogens in dogs (*Ehrlichia canis*, *Babesia* spp and *Hepatozoon canis,* according to Kledmanee *et al.*, 2009). Although not tested in cats, such method might be applicable also for diagnosis of many feline piroplasmida and *H. canis.*

In this chapter a new quantitative PCR assay for the detection of feline piroplasmids and *Hepatozoon* is presented. The method is based on the procedure published by Criado-Fornelio *et al.*, (2009a), previously devised for diagnosis of bovine piroplasmids. In the present work, the target sequence was the 5' end of the 18S rRNA gene of *Cytauxzoon* sp (Spanish cat). Primers UPFR-F = 5'-ttgatcctgccagtagt-3' and UPFR-R=5'-agctgatrggtcagaaac-3 were combined with FRET probes TAFA (5'- actgttgtaatgagccattcgcagtttc-3' Fluorescein) and TAFS (5'LCRED 705-tataaaagcttatacttagacatgcatggc-3'). All assays were performed employing the Lightcycler 1.5 thermocycler (Roche Diagnostics, Manheinn, Germany) following the manufacturer's recommendations. Assay's specificity was tested as previously reported (Criado-Fornelio *et al.*, 2009a), by using DNA samples of *Toxoplasma*, *Neospora* and *Leishmania*-infected animals as negative amplification controls. Positive amplification controls were DNA samples of carnivores infected by *Cytauxzoon* sp, *Theileria annae*, *B. canis canis*, *B. canis vogeli*, *Hepatozoon felis* and *H. canis*), which had been diagnosed as such in previous studies (Criado-Fornelio *et al.*, 2003a and b, 2004 and 2006).

The optimal thermocycling conditions were an initial denaturation cycle at 95ºC for 10 min, followed by 40 amplification cycles (with a temperature transition rate of 20ºC/s in all steps) consisting of 95ºC for 5 s, annealing at 55ºC for 15 s, and an extension step at 72ºC for 15 s. Automated fluorescence reading and recording was set at "single" in the annealing step. The reaction mix contained 4 μl of FastStart DNA MasterPlus Hybprobe (Roche), 0.5 μM of each primer, 2 μl each of sensor and anchor probe (20 μM each), 11 μl of sterile, distilled water, and 2 μl of template DNA (approximately 5-10 ng of DNA per assay).

```
                                   ----------Sensor probe--------      --------Anchor probe-------- / No. differences
AY309956 Cytauxzoon sp             GCCATGCATGTCTAAGTATAAGCTTTTATA-TGGT-GAAACTGCGAATGGCTCATTAAAACAGT   0 = PROBE
AY150068 Theileria annae           GCCATGCATGTCTTAGTATAAGCTTTTATA-CAGC-GAAACTGCGAATGGCTCATTAAAACAGT   1
AY150057 Babesia canis canis       GCCATGCATGTCTAAGTACAAGCTTTTTA -CGGT-GAAACTGCGAATGGCTCATTACAACAGT   4
AY628681 Hepatozoon felis Spain 2  GCCATGCATGTCTAAGTATAAACAGTAATA-CAGT-AAAACTGCAAATGGCTCATTAAAACAGT   6
AY620232 Hepatozoon felis Spain 1  GCCAGGCATGTCTAAGTATAAACAGTAATA-CAGT-AAAACTGCAAATGGCTCATTAAAACAGT   7
AY461378 H. canis Spain 2          GCCATGCATGTCTAAGTATAAACAAAAATA-CAGT-AAAACTGCAAATGGCTCATTACAACAGT   8
                                   **** ******** **** **                ******* *********** ******
```

Figure 2. Alignment of piroplasmid 18S rRNA gene sequences targeted by the FRET probes. Only the hepatozoons and piroplasmids most commonly found in European felines were included in the alignment. The GenBank® accession numbers for piroplasmid sequences, FRET probe binding regions (both anchor and sensor), and the number of base differences (compared to FRET probes) are indicated. Conserved positions in piroplasmid sequences are shown by asterisks. Variable positions in sequences are underlined.

The sensitivity of the FRET assay was tested as previously published by Buling *et al.*, (2007). Serial dilutions of linearized p-GEM plasmid with the *Cytauxzoon* sp DNA insert were used for these performance assays. The new qPCR method showed good specificity, since no amplification products were detected in negative control samples, whereas strong amplification signal was evident in all piroplasmid-positive samples. The amplification curves for *H. felis* and *H. canis* exhibited a relatively weaker amplification signal due to the six or more mismatches of the fluorescence probe (the alignment of sequences is shown in Figure 2), while the melting curve analysis was not hampered by such circumstance.

The Student t test (set at $P<0.001$ or lower) was used to confirm the existence of statistically significant differences between melting peak temperatures of piroplasmid or *Hepatozoon* species. Due to sequence differences in the target region of the probe, the FRET assay readily distinguished between the chosen feline hematozoan species, with two exceptions: species of the *B. canis* complex plus *H. canis* and *H. felis* (Figure 3). In this respect, *H. felis* Spain 1 showed an average melting temperature (MT) of 56,5 ºC. In contrast, *H. felis* Spain 2 and *H. canis* showed average MT of 55,8 ºC and 55,4 ºC, respectively. The difference between those temperatures was non significant when compared by the Student t test. Therefore, it was concluded that the assay detected *Hepatozoon* sp rather than a particular species.

The sensitivity of the proposed technique was about 15 copies of plasmid DNA. The assay provided linear response curves over the 10^3 – 10^8 copies range. The standard curve, determined with the Lightcycler software, showed excelent performance (correlation coefficient = 0.99 and slope -3.2); these data suggested that the qPCR assay had been set up correctly.

The new assay was used for screening 450 blood samples of domestic cats from Spain. The samples were obtained by veterinarians in a local animal shelter and then send for diagnosis to our laboratory with a frozen gel pack. Extraction of DNA from mammalian blood was carried out with the Blood Spin DNA kit (Mobio, Solana Beach, CA, USA). The animals studied were either asymptomatic or showed some signs of disease, usually not related to piroplasmosis. The qPCR technique detected 1 cat positive for *Cytauxzoon* sp, 2 cats positive for *T. annae* and 1 cat positive for *B. canis canis*, whereas standard PCR (Criado-Fornelio *et al.*, 2007) only detected the cat positive for *Cytauxzoon* sp. DNA sequencing showed that all isolates were genetically identical (or very closely related) to those previously reported by us in Portugal or Spain. Two of these piroplasmid-infected cats were positive at the same time for *Hepatozoon* spp. These data suggest that the qPCR assay is more sensitive than standard PCR, as shown in previous studies (Li *et al.*, 2008; Criado-Fornelio *et al.*, 2009a).

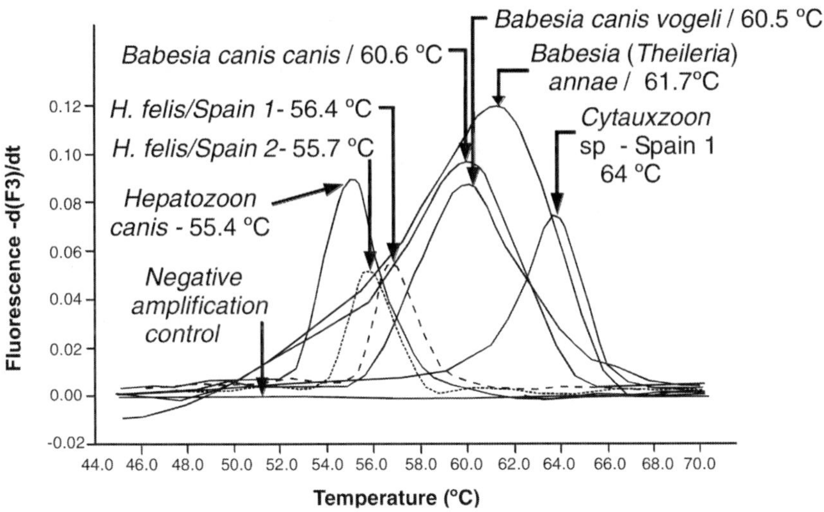

Figure 3. Melting curve analysis obtained in the FRET probe method. Curves obtained with control samples (*Babesia canis canis*, *B. canis vogeli*, *Cytauxzoon* sp, *Babesia (Theileria) annae*, *Hepatozoon canis*, *H. felis* (genotypes Spain 1 and 2) and negative control (*Leishmania infantum*). Peaks not always showed significant differences in melting temperature (when comparing species in pairs - see the chapter text for explanation).

The new FRET assay permitted identification at the genus level only in the *B. canis* complex. This might be seen as a flaw of the method. The problem existing in discrimination of the *B. canis* complex subspecies by the qPCR method has been as well noticed by Wang *et al.*, (2010). Notwithstanding, such failure in species identification may be of little importance in clinical practice, at least in Europe. There are no specific chemotherapy regimens for the different subspecies of the *B. canis* complex (Baneth *et al.*, 1998; Baneth *et al.*, 2004; Solano and Baneth, 2011). In other respects, the discrimination of *Hepatozoon* species was not posible with the new qPCR assay. This possible trouble may be negligible for chemotherapy of cats in Europe. Anyway, there had been no reports of treatment in *H. felis*-infected cats, so that it is not possible to know if the treatment employed for *H. canis* is useful for all feline hepatozoons. Theoretically this new method can not distinguish between *H. felis* and *H. americanum*, because both species show 6 base differences in the probe target region. This may be a problem, since the treatment for *H. canis* differs from that of *H. americanum* (Allen *et al.*, 2010), however the latter species has been seldom found in cats (Allen *et al.*, 2011). If *Hepatozoon* species recognition is essential, the alternative qPCR method described in the next chapter (Emerging tick-borne protozoal diseases - part 2) may be useful. Otherwise DNA sequencing may be employed to identify protozoan isolates (Criado-Fornelio *et al.*, 2003a,b and 2007).

Table 3. Summary of epizootiologic findings related to feline piroplasmosis. Emphasis is given to recent publications where diagnosis was based on molecular methods

Author	Country	Diagnostic method	Host / Piroplasmid species found	Prevalence / No. animals studied
Averbeck *et al.*, (1990)	Tanzania	MIC	Lion and Cheetah / *Theileria sp*	100% and 100% / 123 and 8
Schoeman *et al.*, (2001)	South Africa	MIC	Domestic cat / *B. felis* / NA	NA / 56
Penzhorn *et al.*, (2001)	South Africa	PCR+S	Lion, cat, caracal / *B. leo*, *Babesia* sp, *B. felis*	NA / 18
Criado-Fornelio *et al.*, (2003)	Portugal	PCR+S	Domestic cat / *Babesia* (*Theileria*) *annae* and *B. canis canis*	30% and NA / 16 and 1
Ketz-Riley *et al.*, (2003)	Mongolia	MIC and PCR+S	Manul cat / *Cytauxzoon manul*	100 % / 4
Criado-Fornelio *et al.*, (2004)	Spain	PCR+S	Domestic cat / *Cytauxzoon* sp	1% / 100
Baneth *et al.*, (2004)	Israel	PCR+S	Domestic cat / *B. canis presentii*	NA / 2
Mendes-de-Almeida *et al.*, (2004)	Brazil	MIC	Domestic cat / Non-identified piroplasmids	47% / 47
Soares *et al.*, (2004)	Brazil	MIC	Mountain lion / *Cytauxzoon felis*-like	NA / 1
Bondy *et al.*, (2005)	USA	PCR+S	*Amblyomma americanum* / *C. felis*	< 0.002 % / 1362
Birkenheuer *et al.*, (2006b)	USA	PCR+S	Domestic cat / *C. felis*	NA / 34
Yabsley *et al.*, (2006)	USA	PCR+S	Cougar / *Cytauxzoon felis* and *Babesia* sp	95 and 78 % / 41
Haber *et al.*, (2007)	USA	PCR+S	Domestic cat / *C. felis*	0.3% / 961
Pennisi *et al.*, (2007)	Italy	PCR+S	Domestic cat / *Babesia* (*Theileria*) *annae*	22% / 20

Author	Country	Diagnostic method	Host / Piroplasmid species found	Prevalence / No. animals studied
Carneiro (2007)	Brazil	MIC and PCR+S	Domestic cat / *B. canis vogeli*	MIC = 3.4% / 144 PCR+S = 0.9 % / 110
Bosman *et al.*, (2007)	South Africa	RLB, PCR+S	Lion, cheetah and domestic cat / *Babesia* sp, *B. felis* and *B. leo*	7.5, 50 and 50% / 81, 137 and 27
Millán *et al.*, (2007)	Spain	PCR+S	Iberian lynx / *Cytauxzoon* sp	10% / 20
Georges *et al.*, (2008)	Trinidad	RLB, PCR+S	Domestic cat / *B. canis vogeli*	6.7% / 15
Criado-Fornelio *et al.*, (2009b)	France	PCR+S	Domestic cat / *Cytauxzoon* sp	0.8 % / 116
Millán *et al.*, (2009)	Spain	PCR+S	Iberian lynx and domestic cat / *Cytauxzoon* sp	15 and 0 % / 26 + 53
Widmer (2009)	Brazil	PCR+S	Jaguar and diverse ticks (*Amblyomma cajennense*, *A. triste* and *Riphicephalus microplus)* / *C. felis*	100 and 100% / 10 and 40
Simking *et al.*, (2010)	Thailand	MIC and PCR+S	Domestic cat / *B. canis vogeli*?	0.13 (MIC) and 1.4 % (PCR+S) / 1490
Lledó *et al.*, (2010)	Spain	PCR+S	*Ixodes ricinus* and *I. hexagonus* / *Babesia (Theileria) annae*	NA / 107
Shock *et al.*, (2011)	USA	PCR+S	Bobcat and cougar / *C. felis*	20 % and 14% / 697 and 7
Carli *et al.*, (2011)	Italy	MIC and PCR+S	Domestic cat / *Cytauxzoon* sp	15 and 23 % / 118
André *et al.*, (2011)	Brazil	SER and PCR+S	SER = diverse felids / *Babesia* sp PCR+S = Pampa cat, genet / *B. leo*	10 % / 167 3.7 % / 167
Ahmad *et al.*, (2011)	Pakistan	MIC	Domestic cat / *Babesia* sp	3.1 % / 183
Baraboglia (2011)	Argentina	MIC	Domestic cat / *Cytauxzoon* sp	NA / 5
Filoni *et al.*, (2012)	Brazil	SER; PCR+S	Wild felines / *Cytauxzoon* sp	< 1% / 210

Abbreviations: MIC = diagnosis by microscopy; SER = diverse immunological methods; PCR+S = PCR + sequencing; RLB = Reverse line blotting; NA = Not Applicable (case report, only the animals positive by microscopy were studied, etc).

Finally, it is interesting to mention the existence of an alternative molecular method for detection of feline piroplasmids: the reverse line blotting. Bosman *et al*., (2007) studied the presence of *C, felis*, *B. felis* and *B. leo* in South African felids by this new hybridization assay. Unfortunately, the sequence of the probes used for detection of *C. felis* and *B. leo* were not published, so that other researchers can not check the performance of the assay. Georges *et al*., (2008) studied the presence of piroplasmids in domestic cats from South America (Trinidad) by RLB, finding that one animal was infected by *B. canis vogeli*. In conclusion, this procedure has been seldom employed for diagnosis in felines. This fact may be explained by the need of a specialized laboratory and the relatively high cost of the method for small scale-diagnosis laboratories. However, it must be acknowledged that the method is more flexible and less expensive than DNA microarrays (Kong and Gilbert, 2007).

5. Epizootiology of Feline Piroplasmosis

Table 3 shows a summary of the studies conducted on Epizootiology of feline piroplasms. According to this information, it may be concluded that the distribution of the main feline piroplasms is as follows (Figure 4):

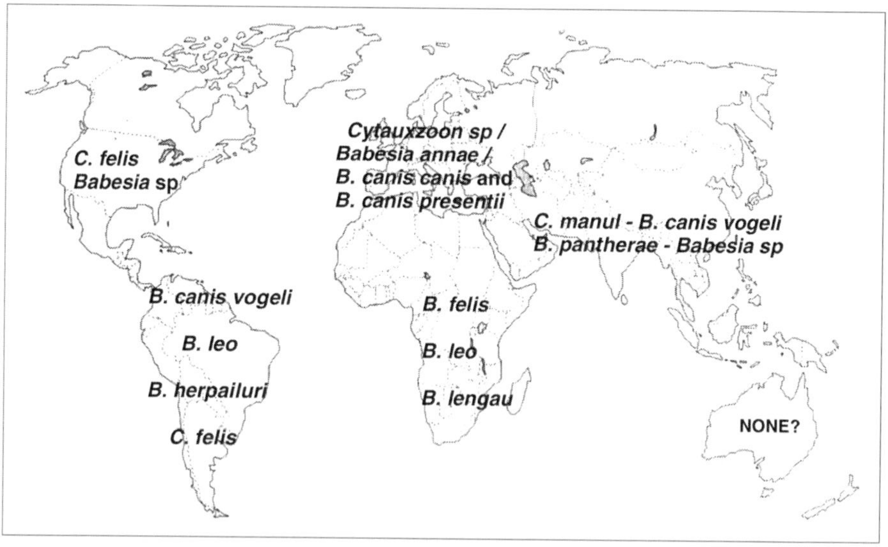

Figure 4. Approximate geographic distribution (by continent) of feline piroplasmids.

a) *Cytauxzoon* species:
   - *Cytauxzoon felis* (*sensu stricto*): The USA and South America. It is important to underline that in Brazil and Argentina, parasite identification has been sometimes based on morphology (Soares, 2001 and Baraboglia, 2011), which makes doubtful the adscription of some South American isolates to this species.
   - *Cytauxzoon sp* ("European strain"): Southern Europe (Spain, Italy, France).
   - *C. manul*: Mongolia.

b) *Babesia* species:
   - *B. herpailuri*: South America, Asia.
   - *B. felis*: South Africa, Sudan?
   - *B.leo*: South Africa and Brazil. The distribution of this species may be broader than expected, since it has been recently diagnosed in wild felines from Brazil by molecular methods (André *et al.*, 2011). Moreover, a genetically related species has been found in baboons in the USA, although it is likely that the parasite could have infected the animals in Africa (Reichard *et al.*, 2011). The authors indicated that protozoa transmission within apes of the same species might have occurred in the USA as well.
   - *B. canis canis/presentii*: possibly Southern Europe and Middle East.
   - *B. canis vogeli* (or related organisms): Far East (Thailand), South America (Brazil and Trinidad). Possibly this subspecies has a cosmopolitan distribution (associated to *Riphicephalus sanguineus*) but its frequency may be quite low in some parts of the world (*e.g.* Europe, as it has also been shown for dogs). Due to the short sequence obtained in the cat isolate from Thailand (Simking *et al.*, 2010) it is not totally sure its adscription to the subspecies *vogeli*. This is an issue in need of further research.
   - *B. lengau*: South Africa.
   - *B. cati*: this species was named based only on morphology (Mudaliar *et al.*, 1950), which raises some doubts about the identification of the isolate. It may well correspond to *B. pantherae, B. herpailuri* or *B. canis vogeli*, which are "large" piro-plasmid species.

- *Babesia (Theileria) annae* (*Babesia microti*-like): Southern Europe (Spain, Italy). The protozoa might be found as well in felids from the USA or Canada, since *Babesia microti*-like isolates were detected in some wildlife species (such as foxes, as demonstrated by Birkenheuer *et al.*, 2010 and Clancey *et al.*, 2010).

The frequency of feline babesiosis is quite irregular (Table 3). In domestic cats from European countries, prevalence ranges from 0 to 20% (Millán *et al.*, 2009; Criado-Fornelio *et al.*, 2003a, 2004, 2009b; Pennisi *et al.*, 2007 and Carli *et al.*, 2012). Such a high variability may be caused by different parasite species, climatic/geographic factors or in some instances to studies on disease outbreaks. Relatively high prevalence figures may also be found in wild felines in non-tropical areas (Millán *et al.,* 2007 and 2009). Likewise, there is a wide range of variation in piroplasmosis frequency in tropical countries. Thus *B. canis vogeli* in domestic cats from Thailand showed an infection rate similar to European countries. However other reports in feline wildlife species in tropical areas show prevalence variations from 3.7 to 100% (André *et al.*, 2011; Widmer, 2009).

In what concerns the vertebrate host range of feline piroplasmids, current data suggest that some of these hematozoans may infect a broad range of host species. Some piroplasmids (*e.g. Cytauxzoon spp*) are able to survive in many feline hosts. In contrast, other piroplasmids can switch among diverse mammalian hosts (*B. leo* and its relatives, as well as *Babesia* (*Theileria*) *annae* and most species of the *B. canis* complex, perhaps excluding *B. canis rossi*). The ability to survive in different hosts is favoured by some biological factors affecting both the pathogen and its arthropod vectors. Arthropods with noticeable host switching ability may facilitate pathogen transfer (Harbison and Clayton, 2011). However, this feature does not ensure an easy transfer of hematozoans between hosts. Thus, the tick *Dermacentor marginatus* has a wide host range (Hopla *et al.*, 1994) and it is the putative vector of *Cytauxzoon* sp in Europe. This piroplasmid has been found in cats and Iberian lynx in Europe, but never in dogs or foxes (Criado-Fornelio *et al.*, 2003a, 2004, 2006, 2007 and 2009b; Millan *et al.*, 2007; Giménez *et al.*, 2009; Dodzdek *et al.*, 2010; Carli *et al.*, 2012). This suggests that *Cytauxzoon* sp in the Old World has a relatively strict host-specificity, not being able to survive in canids. The same might be true for *C. felis* in the USA.

The available information on feline piroplasmid vectors is scarce. *Cytauxzoon felis* was assumed to be transmitted by *Dermacentor* species

(Blouin *et al.*, 1984). However, recent studies suggest that *Amblyomma americanum* may play an important role as definitive host in the USA (Reichard *et al.*, 2009 and 2010). *B. canis canis* and *B. canis presentii* may be transmitted as well by *Dermacentor* species (Criado-Fornelio *et al.*, 2003a; Baneth *et al.*, 2004), whereas *B. canis vogeli* might be transmitted by *Riphicepalus sanguineus* (Simking *et al.*, 2010). *Babesia* (*Theileria*) *annae* in Europe might be transmitted by diverse *Ixodes* species or even *R. sanguineus* (Camacho *et al.*, 2003; Sun *et al.*, 2008; Cassini *et al.*, 2009; Iori *et al.*, 2010). The vectors for feline babesiosis in South Africa and South America are not known (Penzhorn *et al.*, 2004; Dantas-Torres *et al.*, 2008; Widmer, 2009; André *et al.*, 2011). Widmer (2009) conducted a survey in 10 Brazilian jaguars and found that all ticks collected were positive for *C. felis* DNA. The tick species recovered from wild felines were *Amblyomma cajennense*, *A. triste* y *Riphicephalus microplus*. However, as correctly pointed out by the author of that work, the presence of parasite DNA does not demonstrate vectorial capacity. On the other hand, *Riphicephalus sanguineus* is the suspected vector for *B. canis vogeli* in dogs and it might be as well in felines, although reports on the presence of that species in cat have been scarce. Some feline piroplasmid species have been seldom reported and there are no data available on their putative vectors (*e.g. B. herpailuri*, *B. pantherae*, *B. lengau*).

Many authors have concluded that feline piroplasmosis has no specific pattern on what concerns cat age, gender or breed (Stewart *et al.*, 1980; Schoeman *et al.*, 2001; Baneth *et al.*, 2004; Harvey *et al.*, 2007; Baraboglia, 2011; Carli *et al.*, 2012). However, feline babesiosis follows (at least in some instances) a temporal trend. In the USA, cytauxzoonosis in cats has a biphasic seasonal pattern, corresponding to late spring and early fall (Reichard *et al.*, 2008). In that country, more than 2/3 of infected domestic cats were found in low density residential areas, which usually are located near to forested regions. This is in agreement with the idea that animals located in urban habitats are less exposed to tick bites than those living in rural areas (Gregoire *et al.*, 2002; Criado-Fornelio *et al.*, 2009b).

There are many endangered feline species, so that the transmission of piroplasmids from wildlife to domestic cats and *viceversa* is a potential problem. The possibility of such kind of negative interaction has been pointed out already for the American bobcat (Blouin *et al.*, 1984), the Iberian lynx (Millán *et al.*, 2007, 2009) or the Serengeti lion (Penzhorn, 2006). Likewise, the expansion of vector tick range due to global warming, accompanied by multi-pathogen attack, might be the cause of severe mortality in wild felines (Munson *et al.*, 2008). But the idea of explaining changes in parasite prevalence based only on climatic change factors it is too simplistic. Probably the reason behind the worldwide expansion of piroplasmid infections is rather multifactorial (Beugnet and Marié, 2009).

6. Taxonomic Issues in Feline Piroplasmids: The Case of *Babesia (Theileria) Annae*

Babesia (Theileria) annae has been found in European cats (Criado-Fornelio *et al.*, 2003; Pennisi *et al.*, 2007). However, the parasite (*Theileria annae*) had been previously diagnosed and briefly described by Zahler *et al.*, (2000) in a Spanish dog. Unfortunately, the latter authors did not define the species in agreement with the International Zoological Nomenclature Code. In addition, it has been pointed out that due to the absence of lymphocitic stages in vertebrates, the adscription of *T. annae* to the genus *Theileria* is questionable (Goethert and Telford, 2003). Hence, the denomination "*Babesia microti*-like" has been subsequently applied in many studies for the canine/feline "theilerid". Nothwithstanding, the use of such names leads to confusion, since sequences of the 18S rRNA gene of *B. microti*-like isolates from small mammals in Japan only showed 95 - 97% identity to *T. annae* (Saito-Ito *et al.*, 2004; Tsuji *et al.*, 2006). The available evidence suggests that *B. microti*-like / *Babesia (Theileria) annae* is capable to parasitize carnivores and ungulates, but not rodents (Goethert and Telford, 2003; Criado-Fornelio *et al.*, 2003a,b, 2006, 2007 and 2010; Zamoto *et al.*, 2004; Pietrobelli *et al.*, 2007; Pennisi *et al.*, 2007; Giménez *et al.*, 2009, Yeagley *et al.*, 2009; Rar *et al.*, 2010; Clancey *et al.*, 2010; Moretti *et al.*, 2010) and therefore it is an entity different from *B. microti*. It is suggested here that in order to avoid any future identification problem *T. annae* should be named *Babesia annae*, as also pointed out by Clancey *et al.*, (2010).

7. Preventive and Control Measures

There are no commercial vaccines for feline babesiosis (Solano-Gallego and Baneth, 2011). A vaccine is available for *B. canis* in dogs (Intervet International®), but it is unclear if such biological preparation (intended only for canids) might induce protection against feline piroplasmid species such as *B. canis canis* or *B. canis presentii*.

Table 4. Results of chemotherapy in felines infected with piroplasmids

Author	Country	Host / Piroplasmid species	Active drug*	Survival rate / No. of animals
Potgieter (1981)	South Africa	Domestic cat / *Babesia felis*	Primaquine	NA – drug with babesicidal effect
Uilenberg *et al.*, (1987)	The Netherlands	Domestic cat / *Cytauxzoon felis*	Parvaquone	NA - drug with babesicidal effect / 1
Moztel and Wagner (1990)	The USA	Domestic cat / *C. felis*	Parvaquone/ buparvaquone	6% / 15
Hoover *et al.*, (1994)	The USA	Domestic cat / *C. felis*	Ampicillin, doxycycline	0% / 8
Walker and Cobell, (1995)	The USA	Domestic cat / *C. felis*	Doxycycline	NA - drug with babesicidal effect / 1
Greene *et al.*, (1999)	The USA	Domestic cat / *C. felis*	Imidocarb, diminazene aceturate	85 % / 7
Penzhorn *et al.*, (2000)	South Africa	Domestic cat / *Babesia felis*	Rifampin; trimethroprim + Sulphadizine/	NA – Drugs with some babesicidal effect / 15
Meinkoth *et al.*, (2000)	The USA	Domestic cat / *C. felis*	Imidocarb	100% – babesicidal effect not evident / 18 cases studied, but only 1 cat was treated
Baneth *et al.*, (2004)	Israel		Imidocarb, Doxycycline	100 % / 2
Brown *et al.*, (2009)	The USA	Domestic cat / *C. felis*	Imidocarb	100 % / 2
Carli *et al.*, (2011)	Italy	Domestic cat / *Cytauxzoon* sp	Imidocarb, doxycycline azythromycin, atovaquone (+ immunosupresive treatment)	33% / 3

Table 4. (Continued)

Author	Country	Host / Piroplasmid species	Active drug*	Survival rate / No. of animals
Cohn *et al*., (2011)	The USA	Domestic cat / *C. felis*	Imidocarb Azythromycin + atovaquone 60% / 32	27% / 53 (imidocarb) 60% / 32 (Azyth. + Atov.)

Abbreviations : NA = not applicable (individual case report, no reported deaths or survival rate not indicated).

* = in some instances several drugs were employed in chemotherapy. However, for simplicity, only the drug finally recommended by the authors is indicated.

As is for other tick-borne diseases, the prevention of feline babesiosis is not an easy task. Attempts to reduce piroplasmosis prevalence are based on avoiding tick attachment by means of chemical repellents or alternatively killing these arthropods with acaricides (Dryden, 2009). Chemical attack may be applied with many procedures, such as pipetes, collars, bathing, spraying, dipping or dusting (Shaw and Day, 2005). Some products like permethrin are both repelent and acaricide at the same time. DEET (N,N-diethyl-toluamide) is used as insect or tick repellent but some studies disagree on its efficacy against ticks (Miller *et al.*, 2011; Bissinger *et al.*, 2011). In addition, environmental control measures may be undertaken (Blagburn and Dryden, 2009). To alter tick habitat, activities such as removing leaves and clearing brush and tall grass from around the house can help reduce the number of ticks. Because rodents, deer, and other animals can harbour ticks, it is advisable to limit the presence of these animals (near to human habitat) as far as possible. Some authors have indicated that euthanasia of stray cats and dogs improves the results of environmental treatment with acaricides (Jurisic *et al.*, 2010), but such strategy should be avoided if rescuing cats in animal shelters is feasible. However, animal rescue activity does not ensure that abandoned pets are spared. It has been shown that economic constraints may increase the number of euthanized animals in shelters (Weng and Hart, 2012).

Due to the sporadic nature of feline babesiosis, chemotherapeutic trials employing significant number of cats have been scarce (Table 4). Therefore the treatment of feline piroplamosis is quite controversial. Some authors find effective a treatment regime but the same protocol may completely fail when it is applied by other researchers. A likely explanation for the controversy is that in these chemotherapeutic assays, drugs were administered at different moments of evolution of the disease. In addition, inconsistencies in the treatment outcome may have been caused by the existence of different parasite strains infecting animals in different locations (Meinkoth *et al.*, 2000). To further complicate this issue, there might have been some concomitant viral or bacterial infections which precluded a satisfactory chemotherapeutic action.

Treatments with both berenil and imidocarb seemed to be effective against *B. herpailuri*, as shown by detailed ultrastructural studies by Hebel and Dennig, (1969) and Gobel and Dennig, (1976). Primaquine seems to be effective for *B. felis* in cats (Potgieter, 1981). Diminazene aceturate or imidocarb have been used with success for feline cytauxzoonosis (Greene *et al.*, 1999; Brown *et al.*, 2008). In some instances, the treatment with antibiotics has led to the total recovery of diseased cats (Hoover *et al.*, 1994). However, some authors found no effect when applying antibiotic therapy (Walker and Cobell, 1995). Recently, a therapy trial using a combination of atovaquone and azithromycin (Cohn *et al.*, 2011) was effective in curing one group of 32 *Cytauxzoon*-infected cats in the USA. These results were superior to that obtained with the standard imidocarb therapy. The only possible constraint for this new approach is the relatively high price of such a treatment. However, in the long run, the issue of cost should not influence the choice of atovaquone and azithromycin for treatment, as pointed out in humans (Krause *et al.*, 2000).

CONCLUSION

Feline piroplasmosis may be considered an emergent disease. Many new hematozoans have been discovered since the arrival of molecular diagnosis methods. The presence of such a diverse parasitic fauna in felines should stress the need of further studies on both arthropod vectors and reservoir hosts. Special attention should be paid to piroplasmid taxonomy, to avoid confusion in organism identification, as now is the case of *Babesia* (*Theileria*) *annae*. However, the problem is quite complicated, especially if the status of some species is to be reconsidered.

In spite of the fact that up to now there had been no attempts to develop vaccines for feline piroplasmosis, the diversity of potentially pathogenic species may affect issues such as immunotherapy and chemotherapy of the disease. Thus, intra- and inter-specific antigen diversity may affect vaccine efficacy. The existence of genetic variability in hematozoans may also help to explain the controversial results obtained in chemotherapy.

Most piroplasmid-infected felines are asymptomatic carriers. However, the existence of concurrent infections may worsen disease signs and complicate prognosis. Such a combined pathogenic attack may be associated with changes in multiple factors. Among them, those derived from global warming have received a lot of attention from epidemiologists and parasitologists. It seems that variations in climatic conditions affect the distribution of pathogens and their vectors and therefore represent a problem for endangered feline species.

All these facts underline the need of focusing research efforts in the coming years on issues such as diagnosis, epizootiology and chemotherapy of piroplasmosis in felids.

ACKNOWLEDGMENTS

I am deeply indebted to Dr. Jeronimo Ruas for his critical reading of the manuscript. I also wish to thank the veterinarians of the SMP animal shelter for their invaluable help in the collection of animal samples.

REFERENCES

Ahmad, S.S., Khan, M. S., Khan, M. A., Ahmad, N., (2011). Prevalence of babesiosis in cats in Lahore, Pakistan. *The Journal of Animal and Plant Sciences,* 21, 354-357.

Allen, K., Little, S., Johnson, E., Hostetler, J., Panciera, R., Ewing, S., (2010). Treatment of *Hepatozoon americanum* Infection: Review of the literature and experimental evaluation of efficacy. *Veterinary Therapeutics*, 11, E1-E8.

Allen, K.E., Yabsley, M.J., Johnson, E.M., Reichard, M.V, Panciera, R.J., Ewing, S.A., Little, S.E., (2011). Novel *Hepatozoon* in vertebrates from the southern United States. *The Journal of Parasitology,* 97, 648-653.

Allsopp, M.T., Cavalier-Smith, T., De Waal, D.T., Allsopp, B.A., (1994). Phylogeny and evolution of the piroplasms. Parasitology, 108, 147-52.

André, M.R., Adania, C.H., Rodrigo Teixeira, H.F., Allegretti, S.M., Machado, R.Z., (2011). Molecular and serological detection of *Babesia* spp. in neotropical and exotic carnivores in Brazilian Zoos. *Journal of Zoo and Wildlife Medicine,* 42, 139-143.

Ano, H., Makimura, S., Harasawa, R., (2001). Detection of *Babesia* species from infected dog blood by polymerase chain reaction. *Journal of Veterinary Medical Science,* 63, 111-113.

Averbeck, G.A., Bjork, K.E., Packer, C., Herbst, L., (1990). Prevalence of hematozoans in lions (*Panthera leo*) and cheetah (*Acinonyx jubatus*) in Serengeti National Park and Ngorongoro Crater, Tanzania. *The Journal of Wildlife Diseases,* 26, 3942-3944.

Baneth, G., Aroch, I., Tal, N., Harrus, S., (1998). *Hepatozoon* species infection in domestic cats: a retrospective study. *Veterinary Parasitology,* 79,123-133.

Baneth, G., Kewnny, M.J., Tasker, S., Anug, Y., Shkap, V., Levy, A., Shaw, S.E., (2004). Infection with a proposed new subspecis of *Babesia canis, Babesia canis* subsp. *presentii,* in domestic cats. *Journal of Clinical Microbiology,* 42, 99-105.

Baraboglia, E.R., (2011). Hallazgo de un hemoparásito eritrocitico tipo *Cytauxzoon felis* en *Felis catus domesticus, Redvet,* 12, 1-29.

Beugnet, F., Marié, J.L., (2009). Emerging arthropod-borne diseases of companion animals in Europe. Veterinary Parasitology, 163, 298-305.

Birkenheuer, A.J., Marr, H., Alleman, A.R., Levy, M.G., Breitschwerdt, E.B., (2006a). Development and evaluation of a PCR assay for the detection of *Cytauxzoon felis* DNA in feline blood samples. *Veterinary Parasitology,* 137, 144-9.

Birkenheuer, A.J., Le, J.A., Valenzisi, A.M., Tucker, M.D., Levy, M.G., Breitschwerdt, E.B., (2006b). *Cytauxzoon felis* infection in cats in the mid-Atlantic states: 34 cases (1998-2004). *Journal of the American Veterinary Medical Association,* 228, 568-571.

Birkenheuer, A.J., Horney, B., Bailey, M., Scott, M., Sherbert, B., Catto, V., Marr, H.S., Camacho, A.T., Ballman, A.E., (2010). *Babesia microti*-like infections are prevalent in North American foxes. *Veterinary Parasitology,* 172, 179-182.

Blagburn, B.L, Dryden, M.W., (2009). Biology, treatment, and control of flea and tick infestations. *Veterinary Clinics North American Small Animals Practitioner,* 39,1173-1200.

Bissinger, B.W., Apperson, C.S., Watson, D.W., Arellano, C., Sonenshine, D.E., Roe, R.M., (2011). Novel field assays and the comparative repellency of BioUD(®), DEET and permethrin against *Amblyomma americanum. Medical and Veterinary Entomology,* 25, 217-26.

Blouin, E.F., Kocan, A.A., Glenn, B.L., Kocan, K.M., Hair, J.A., (1984). Transmission of *Cytauxzoon felis* Kier, 1979 from bobcats, *Felis rufus* (Schreber), to domestic cats by *Dermacentor variabilis* (Say). *The Journal of Wildlife Diseases,* 20, 241-2.

Böhm, M., Leisewitz, A.L., Thompson, P.N., Schoeman, J.P., (2006). Capillary and venous *Babesia canis rossi* parasitaemias and their association with outcome of infection and circulatory compromise. *Veterinary Parasitology,* 141, 18- 29.

Bondy, P.J., Cohn, L.A., Tyler, J.W., Marsh, A.E., (2005). Polymerase chain reaction detection of *Cytauxzoon felis* from field-collected ticks and sequence analysis of the small subunit and internal transcribed spacer 1 region of the ribosomal RNA gene. *The Journal of Parasitology,* 91, 458-461.

Bosman, A.M., Venter, E.H., Penzhorn, B.L., (2007). Occurrence of *Babesia felis* and *Babesia leo* in various wild felid species and domestic cats in Southern Africa, based on reverse line blot analysis. *Veterinary Parasitology,* 144, 33-38.

Bosman, A.M., Oosthuizen, M.C., Peirce, M.A., Venter, E.H., Penzhorn, B.L., (2010). *Babesia lengau* sp. nov., a novel *Babesia* species in cheetah (*Acinonyx jubatus*, Schreber, 1775) populations in South Africa. *Journal of Clinical Microbiology, 48*, 2703-2708.

Brown, H.M., Latimer, K.S., Erikson, L.E., Cashwell, M.E., Britt, J.O., Peterson, D.S., (2008). Detection of persistent *Cytauxzoon felis* infection by polymerase chain reaction in three asymptomatic domestic cats. *Journal of Veterinary Diagnostic Investigations,* 20, 485-8.

Brown, H.M., Berghaus, R.D., Latimer, K.S., Britt, J.O., Rakich, P.M., Peterson, D.S., (2009). Genetic variability of *Cytauxzoon felis* from 88 infected domestic cats in Arkansas and Georgia. *Journal of Veterinary Diagnostic Investigations,* 21, 59-63.

Buling, A., Criado-Fornelio, A., Asenzo, G., Benitez, D., Barba-Carretero, J.C., Florin-Christensen, M., (2007). A quantitative PCR assay for the detection and quantification of *Babesia bovis* and *B. bigemina*. *Veterinary Parasitology,* 147, 16-25.

Camacho, A.T., Pallas, E., Gestal, J.J., Guitián, F.J., Olmeda, A.S., Goethert, H.K., Telford, S.R., (2001). Infection of dogs in north-west Spain with a *Babesia microti*-like agent. *The Veterinary Record,* 149, 552-555.

Camacho, A.T., Pallas, E., Gestal, J., Guitián, J., Olmeda, A., Kenny, M., Telford, S., Spielman, A., (2002). *Babesia microti*: ¿una nueva forma de babesiosis humana en Europa? *Enfermedades Infecciosas y Microbiología Clínica,* 20, 417-418.

Camacho, A.T., Pallas, E., Gestal, J., Guitián, J., Olmeda, A., Telford, S.R., Spielman, A., (2003). *Ixodes hexagonus* is the main candidate as vector of *Theileria annae* in northwest Spain. *Veterinary Parasitology,* 112, 157-163.

Carli, E., Trotta, M., Chinelli, R., Drigo, M., Sinigoi, L., Tosolini, P., Furlanello, T., Millotti, A., Caldin, M., Solano-Gallego L. (2012). *Cytauxzoon* sp. infection in the first endemic focus described in domestic cats in Europe. *Veterinary Parasitology*, 183, 343-352.

Carneiro, M.P.M., (2007). Ocorrência de infecções por *Babesia* spp e *Hepatozoon* spp em gatos domésticos (*Felis domesticus*) do estado de São Paulo e do distrito federal. *Master Dissertation,* College of Veterinary Medicine and Zootechny, Universidade Estadual Paulista (Sao Paulo, Brazil), pp. 1-71.

Carpano, M., (1934). *Sur les piroplasmoses des carnassiers et sur un nouveau piroplasme des felins* (*Babesiella felis* chez le puma: *Felis concor*). Bulletin, Ministry of Agriculture, Egypt - Technical and Scientific Service, No. 136. 20 pp.

Cassini, R., Zanutto, S., Frangipane-di-Regalbono, A., Gabrielli, S., Calderin, P., Moretti, A., Tampieri, M.P., Pietrobelli, M., (2009). Canine piroplasmosis in Italy: epidemiological aspects in vertebrate and invertebrate hosts. *Veterinary Parasitology,* 165, 30-35.

Clancey N, Horney B, Burton S, Birkenheuer A, McBurney S, Tefft K. *Babesia* (*Theileria*) *annae* in a red fox (*Vulpes vulpes*) from Prince Edward Island, Canada. *Journal of Wildlife Diseases,* 46, 615-621.

Cowell, R.L., Fox, J.C., Panciera, R.J., Tyler, R.D., (1988). Detection of anticytauxzoon antibodies in cats infected with a *Cytauxzoon* organism from bobcats. *Veterinary Parasitology,* 28, 43-52.

Cohn, L.A., Birkenheuer, A.J., Brunker, J.D., Ratcliff, E.R., Craig, A.W., (2011). Efficacy of atovaquone and azithromycin or imidocarb dipropionate in cats with acute cytauxzoonosis. *Journal of Veterinary Internal Medicine,* 25, 55-60.

Criado-Fornelio, A., Martinez-Marcos, A., Buling-Saraña, A., Barba-Carretero, J.C., (2003a). Presence of *Mycoplasma haemofelis*, *Mycoplasma haemominutum* and piroplasmids in cats from southern Europe: a molecular study. *Veterinary Microbiology,* 93, 307-317.

Criado-Fornelio, A., Martinez-Marcos, A., Buling-Saraña, A., Barba-Carretero, J.C., (2003b). Molecular studies on *Babesia*, *Theileria* and *Hepatozoon* in southern Europe. Part II. Phylogenetic analysis and evolutionary history. *Veterinary Parasitology,* 114, 173-194.

Criado-Fornelio, A., Gónzalez-del-Río, M.A., Buling-Saraña, A., Barba-Carretero, J.C., (2004). The "expanding universe" of piroplasms. *Veterinary Parasitology,* 119, 337-345.

Criado-Fornelio, A., Ruas, J.L., Casado, N., Farias, N.A., Soares, M.P., Müller, G., Brumt, J.G., Berne, M.E., Buling-Saraña, A., Barba-Carretero, J.C., (2006). New molecular data on mammalian *Hepatozoon* species (Apicomplexa: Adeleorina) from Brazil and Spain. *Journal of Parasitology,* 92, 93-99.

Criado-Fornelio, A., Rey-Valeiron, C., Buling, A., Barba-Carretero, J.C., Jefferies, R., Irwin, P., (2007). New advances in molecular epizootiology of canine hematic protozoa from Venezuela, Thailand and Spain. *Veterinary Parasitology,* 144, 261-269.

Criado-Fornelio, A., Buling, A., Asenzo, G., Benitez, D., Florin-Christensen, M., Gonzalez-Oliva, A., Henriques, G., Silva, M., Alongi, A., Agnone, A., Torina, A., Madruga, C.R., (2009a). Development of fluorogenic probe-based PCR assays for the detection and quantification of bovine piroplasmids. *Veterinary Parasitology,* 162, 200-206.

Criado-Fornelio, A., Buling, A., Pingret, J.L., Etievant, M., Boucraut-Baralon, C., Alongi, A., Agnone, A., Torina, A., (2009b). Hemoprotozoa of domestic animals in France: prevalence and molecular characterization. *Veterinary Parasitology.* 159, 73-76.

Dantas-Torres F., (2008). *Canine vector-borne diseases in Brazil. Parasites and Vectors,* 1: 25 (17 p).

Davis, L.J., (1929). On a piroplasm of Sudanese wild cat (*Felis ocreatus*). *Transactions of the Royal Society of Tropical Medicine and Hygiene,* 22, 523-534.

Dennig, H.K., (1967). Eine unbekannte Babesienart beim Jaguarundi (*Herpailurus yaguarundi*). *Die Kleintierpraxis,* 12, 146–152.

Dennig, H.K., Brockelsby, D.W., (1972). *Babesia pantherae* sp. nov., a piroplasm of the leopard (*Pantera pardus*). Parasitology, 64, 525–532.

Dezdek, D., Vojta, L., Curković, S., Lipej, Z., Mihaljević, Z., Cvetnić, Z., Beck, R., (2010). Molecular detection of *Theileria annae* and *Hepatozoon canis* in foxes (*Vulpes vulpes*) in Croatia. *Veterinary Parasitology.* 172, 333-336.

Dryden, M.W., (2009). Flea and tick control in the 21st century: challenges and opportunities. *Veterinary Dermatology,* 20, 435-40.

Filoni, C., Catão-Dias, J.L., Cattori, V., Willi, B., Meli, M.L., Ramiro-Corrêa, S.H., Cristina-Marques, M., Harumi-Adania, C., Ramos-Silva, J.C., Vianna-Marvulo, M.F., Ferreira-Neto, J.S., Luiz-Durigon, E., de Carvalho, V.M., Dall'acqua-Coutinho, S., Lutz, H., Hofmann-Lehmann, R., (2012). Surveillance using serological and molecular methods for the detection of infectious agents in captive Brazilian neotropic and exotic felids. *Journal of Veterinary Diagnostic Investigations,* 24, 166-173.

Futter, G.J., Belonje, P.C., (1980a). Studies on feline babesiosis. 2. Clinical observations. *Journal of the South African Veterinary Association,* 51, 143-146.

Futter, G.J., Belonje, P.C., Van den Berg, A., (1980b). Studies of feline babesiosis 3. Haematological findings. *Journal of the South African Veterinary Association,* 51, 271-280.

Georges, K., Ezeokoli, C.D., Newaj-Fyzul, A., Campbell, M., Mootoo, N., Mutani, A., Sparagano, O.A., (2008). The application of PCR and reverse line blot hybridization to detect arthropod-borne hemopathogens of dogs and cats in Trinidad. *Annals of the New York Academy of Sciences,* 1149, 196-199.

Giménez, C., Casado, N., Criado-Fornelio, A., de Miguel, F.A., Domínguez-Peñafiel, G., (2009). A molecular survey for Piroplasmida and *Hepatoozon* in wild animals in Burgos (Northern Spain). *Veterinary Parasitology,* 162, 147-150.

Göbel, E., Dennig, H.K., (1976). The ultrastructure of the intraerythrocytic stages of *Babesia herpailuri*-trophozoites and merozoites after treatment with imidocarb (3,3'-bis-2-imidazolin-2-yl)-carbanilide. *Zeitschrift für Parasitenkunde,* 49, 97-112.

Goethert, H.K., Telford, R., (2003). What is *Babesia microti*? *Parasitology,* 127, 301-309.

Greene, C.E., Latimer, K., Hopper, E., Shoeffler, G., Lower, K., Cullens, F., (1999). Administration of diminazene aceturate or imidocarb dipropionate for treatment of cytauxzoonosis in cats. *Journal of the American Veterinary Medical Association,* 215, 497-500.

Gregoire, A., Faivre, B. Heeb, P., Cezilly, F., 2002. A comparison of infestation patterns by *Ixodes* ticks in urban and rural populations of the Common Blackbird *Turdus merula*. *Ibis,* 144, 640-645.

Gray, A.D., Akin, M.T., McLean, R., Valeri, C.R., (1991). Evaluation of the Quantitative Buffy Coat (QBC) method to detect malaria-infected red blood cells. *Milestones in Medicine,* 156, 241-245.

Haber, M.D., Tucker, M.D., Marr, H.S., Levy, J.K., Burgess, J., Lappin, M.R., Birkenheuer, A.J., (2007). The detection of *Cytauxzoon felis* in apparently healthy free-roaming cats in the USA. *Veterinary Parasitology,* 146, 316-320.

Harbison, C.W., Clayton, D.H. Community interactions govern host-switching with implications for host-parasite coevolutionary history. *Proceedings of the National Academy of Sciences of the USA,* 108, 9525-9529.

Harvey, J.W., Dunbar, M.R., Norton, T.M., Yabsley, M.J., (2007). Laboratory findings in acute *Cytauxzoon felis* infection in cougars (*Puma concolor couguar*) in Florida. *Journal of Zoo and Wildlife Medicine,* 38, 285-91.

Hebel, R., Dennig, H.K., (1969). Light and electron microscope investigations on *Babesia herpailuri* after treatment with 4-4'-diamidino-diazoaminobenzene (Berenil). *Zeitschrift für Parasitenkunde,* 33, 1-20.

Holman, P.J., Snowden, K.F., (2010). Canine hepatozoonosis and babesiosis, and feline cytauxzoonosis. *Veterinary Clinics of North America: Small Animal Practice,* 39, 1035-1053.

Hoover, J.P., Walker, D.B., Hedges, J.D., (1994). Cytauxzoonosis in cats: eight cases (1985-1992). *Journal of the American Veterinary Medical Association,* 205, 455-60.

Hopla, C.A, Durden, L.A., Keirans, J.E., (1994). Ectoparasites and classification. *Office Internationale des épizooties,* 13, 985-1017.

Iori, A., Gabrielli, S., Calderini, P., Moretti, A., Pietrobelli, M., Tampieri, M.P., Galuppi, R., Cancrini, G., (2010). Tick reservoirs for piroplasms in central and northern Italy. *Veterinary Parasitology,* 170, 291-296.

Jurisic, A.D., Petrovic, A.P., Rajkovic, D.V., Nicin, S.D., (2010). The application of lambda-cyhalothrin in tick control. *Experimental and Applied Acarology,* 52, 101-109.

Ketz-Riley, C.J., Reichard, M.V., Van den Bussche, R.A., Hoover, J.P., Meinkoth, J., Kocan, A.A., (2003). An intraerythrocytic small piroplasm in wild-caught Pallas's cats (Otocolobus manul) from Mongolia. *The Journal of Wildlife Diseases,* 39, 424-430.

Kier, A.B., Wagner, J.E., Kinden, D.A., (1987). The pathology of experimental cytauxzoonosis. *Journal of Comparative Pathology,* 97, 415-32.

Kledmanee, K., Suwanpakdee, S., Krajangwong, S., Chatsiriwech, J., Suksai, P., Suwannachat, P., Sariya, L., Buddhirongawatr, R., Charoonrut, P., Chaichoun, K., (2009). Development of multiplex polymerase chain reaction for detection of *Ehrlichia canis*, *Babesia* spp and *Hepatozoon canis* in canine blood. *Southeast Asian Journal of Tropical Medicine and Public Health,* 40, 35-9.

Kocan, A.A., Blouin, E.F., Glenn, B.L., (1985). Hematologic and serum chemical values for free-ranging bobcats, *Felis rufus* (Schreber), with reference to animals with natural infections of *Cytauxzoon felis* Kier, 1979. *The Journal of Wildlife Diseases, 21*, 190-192.

Kocan, A.A., Kocan, K.M., Blouin, E.F., Mukolwe, S.W., (1992). A redescription of schizogony of *Cytauxzoon felis* in the domestic cat. *Annals of the New York Academy of Sciences, 653*, 161-167.

Kong, F., Gilbert, G., (2007). Multiplex PCR-based reverse line blot hybridization assay (mPCR/RLB) — a practical epidemiological and diagnostic tool. *Nature Protocols* 1, 2668 – 2680.

Krause, P.J., Lepore, T., Sikand, V.K., Gadbaw, J., Burke, G., Telford, S.R., Brassard, P., Pearl, D., Azlanzadeh, J., Christianson, D., McGrath, D., Spielman, A., (2000). Atovaquone and azithromycin for the treatment of babesiosis. *New England Journal of Medicine*, 343, 1454-1458.

Kumar, M., Shekhar, P., Haque, S., Mahto, D., (2008). Feline Babesiosis. Veterinary World, 1, 120-121.

Li, Y., Wang, C., Allen, K.E., Little, S.E., Ahluwalia, S.K., Gao, D., Macintire, D.K., Blagburn, B.L., Kaltenboeck, B., (2008). Diagnosis of canine *Hepatozoon* spp. infection by quantitative PCR. *Veterinary Parasitology,* 157, 50-58.

Lledó, L., Giménez-Pardo, C., Domínguez-Peñafiel, G., Sousa, R., Gegúndez, M.I., Casado, N., Criado-Fornelio, A., (2010). Molecular detection of hemoprotozoa and *Rickettsia* species in arthropods collected from wild animals in the Burgos Province, Spain. *Vector Borne and Zoonotic Diseases,* 10, 735–738.

Lopez-Rebollar LM, Penzhorn BL, de Waal DT, Lewis BD., (1999). A possible new piroplasm in lions from the Republic of South Africa. *Journal of Wildlife Diseases,* 35, 82-85.

Meinkoth, J., Kocan, A.A., Whitworth, L., Murphy, G., Fox, J.C., Woods, J.P., (2000). Cats surviving natural infection with *Cytauxzoon felis*: 18 cases (1997-1998). *Journal of Veterinary Internal Medicine,* 14, 521-525.

Meinkoth, J., Kocan, A.A., (2005). Feline cytauxzoonosis. *Veterinary Clinics of North America: Small Animal Practice*, 35, 89-101.

Mendes-de-Almeida, F., Faria, M.C., Branco, A.S., Serrão, M.L., Souza, A.M., Almosny, N., Charme, M., Labarthe, N., (2004). Sanitary conditions of a colony of urban feral cats (*Felis catus* Linnaeus, 1758) in a zoological garden of Rio de Janeiro, Brazil. *Revista do Instituto de Medicina Tropical de São Paulo*, 46, 269-274.

Millán, J., Naranjo, V., Rodríguez, A., de la Lastra, J.M., Mangold, A.J., de la Fuente, J., (2007). Prevalence of infection and 18S rRNA gene sequences of *Cytauxzoon* species in Iberian lynx (*Lynx pardinus*) in Spain. *Parasitology,* 134, 995-1001.

Millán, J., Candela, M.G., Palomares, F., Cubero, M.J., Rodríguez, A., Barral, M, de la Fuente, J., Almería, S., León-Vizcaíno, L., (2009). Disease threats to the endangered Iberian lynx (*Lynx pardinus*). *The Veterinary Journal,* 182, 114-124.

Miller, N.J., Rainone, E.E., Dyer, M.C., González, M.L., Mather, T.N., (2011). Tick bite protection with permethrin-treated summer-weight clothing. *Journal of Medical Entomology,* 48, 327-33.

Moretti, A., Mangili, V., Salvatori, R., Maresca, C., Scoccia, E., Torina, A., Moretta, I., Gabrielli, S., Tampieri, M.p,. Pietrobelli, M. (2010). Prevalence and diagnosis of *Babesia* and *Theileria* infections in horses in Italy: a preliminary study. *The Veterinary Journal,* 184, 346-350.

Moztel, S.L., Wagner, J.E., (1990). Treatment of experimentally induced cytauxzoonosis in cats with parvaquone and buparvaquione. *Veterinary Parsaitology,* 35, 131-138.

Mudaliar, S. V., Achary, G.R., Alwar, V.S., (1950). On a species of *Babesia* in an Indian wild cat (*Felis catus*). Indian Veterinary Journal 26, 391–395.

Munson, L., Terio, K.A., Kock, R., Mlengeya, T., Roelke, M.E., Dubovi, E., Summers, B., Sinclair, A.R., Packer, C., (2008). Climate extremes promote fatal co-infections during canine distemper epidemics in African lions. *PLoS One,* 3(6):e2545.

Ortuño, A., Castellà, J., Criado-Fornelio, A., Buling, A., Barba-Carretero, J.C., (2008). Molecular detection of a *Hepatozoon* species in stray cats from a feline colony in North-eastern Spain. *Veterinary Journal*, 177, 134-5.

Pennisi, M.G., Alongi, A., Agnone, A., Vitale, F., Reale, S., Torina, A. (2007). Cats as reservoir of *Babesia microti*-like. Parassitologia, 49S1, 100.

Penzhorn, B.L., Lewis, B.D., López-Rebollar, L.M., Swan, G.E., (2000). Screening of five drugs for efficacy against *Babesia felis* in experimentally infected cats. *Journal of the South African Veterinary Association,* 71, 53-7.

Penzhorn, B. L., Kjemtrup, A. M., Lopez-Rebollar, L.M., Conrad, P.A., (2001). *Babesia leo* n. sp. from lions in the Kruger National Park, South Africa, and its relation to other small piroplasms. *Journal of Parasitology,* 87, 681–685.

Penzhorn, B.L., Schoeman, T., Jacobson, L.S., (2004). Feline babesiosis in South Africa: a review. *Annals of the New York Academy of Sciences,* 1026, 183-186.

Penzhorn, B. (2006). Babesiosis of wild carnivores and ungulates. *Veterinary Parasitology,* 138, 11-21.

Pietrobelli, M., Cancrini, G., Moretti, A., Tampieri, M.P., (2007). Animal babesiosis: an emerging zoonosis also in Italy? Parassitologia, 49S1:33-8.

Potgieter, F.T., (1981). Chemotherapy of *Babesia felis* infection: efficacy of certain drugs. *Journal of the South African Veterinary Association,* 52, 289-293.

Rar, V.A., Epikhina, T.I., Livanova, N.N., Panov, V.V., Pukhovskaya, N.M., Vysochina, N.P., Ivanov, I.,I., (2010). Detection of *Babesia* DNA in small mammals and Ixodid ticks in the North Urals, Western Siberia and Far East of Russia. Molecular genetics, *Microbiology and Virology,*25, 118-123.

Reichard, M.V., Van Den Bussche, R.A., Meinkoth, J.H., Hoover, J.P., Kocan, A.A., (2005). A new species of *Cytauxzoon* from Pallas' cats caught in Mongolia and comments on the systematics and taxonomy of piroplasmids. *The Journal of Parasitology,* 91, 420-426.

Reichard, M.V., Baum, K.A., Cadenhead, S.C., Snider, T.A., (2008). Temporal occurrence and environmental risk factors associated with cytauxzoonosis in domestic cats. *Veterinary Parasitology,* 152, 314-320.

Reichard, M.V., Meinkoth, J.H., Edwards, A.C., Snider, T.A., Kocan, K.M., Blouin, E.F., Little, S.E., (2009). Transmission of *Cytauxzoon felis* to a domestic cat by *Amblyomma americanum. Veterinary Parasitology,* 161, 110-115.

Reichard, M.V., Edwards, A.C., Meinkoth, J.H., Snider, T.A., Meinkoth, K.R., Heinz, R.E., Little, S.E., (2010). Confirmation of *Amblyomma americanum* (Acari: Ixodidae) as a vector for *Cytauxzoon felis* (Piroplasmorida: Theileriidae) to domestic cats. *Journal of Medical Entomology,* 47, 890-896.

Reichard. M.V., Gray, K.M., Van den Bussche, R.A., d'Offay, J.M., White, G.L., Simecka, C.M., Wolf, R.F., (2011). Detection and experimental transmission of a novel *Babesia* isolate in captive olive baboons (*Papio cynocephalus anubis*). *Journal of the American Association of Laboratory Animal Sciences,* 50, 500-506.

Rotstein, D.S., Taylor, S.K., Harvey, J.W., Bean, J. (1999). Hematologic effects of cytauxzoonosis in Florida panthers and Texas cougars in Florida. *Journal of Wildlife Diseases,* 35, 613-617.

Saito-Ito, A., Yano, Y., Dantrakool, A., Hashimoto, T., Takada, N., (2004). Survey of rodents and ticks in human babesiosis emergence area in Japan: first detection of *Babesia microti*-like parasites in *Ixodes ovatus. The Journal of Clinical Microbiology*, 42, 2268-70.

Schoeman, T., Lobetti, R.G., Jacobson, L.S., Penzhorn, B.L., (2001). Feline babesiosis: signalment, clinical pathology and concurrent infections. *Journal of the South African Veterinary Association,* 72, 4-11.

Schmidt, G.D., Roberts, L.S. (1981). *Foundations of Parasitology,* 2nd Ed., C.V. Mosby Co. Eds., St Louis, Missouri, USA.

Shock, B.C., Murphy, S.M., Patton, L.L., Shock, P.M., Olfenbuttel, C., Beringer, J., Prange, S., Grove, D.M., Peek, M., Butfiloski, J.W., Hughes, D.W., Lockhart, J.M., Bevins, S.N., VandeWoude, S., Crooks, K.R., Nettles, V.F., Brown, H.M., Peterson, D.S., Yabsley, M.J., (2011). Distribution and prevalence of *Cytauxzoon felis* in bobcats (*Lynx rufus*), the natural reservoir, and other wild felids in thirteen states. *Veterinary Parasitology,* 175, 325-330.

Shaw, S.E., Day, M.J. (2005). *Arthropod-borne infectious diseases of the dog and cat* Ed. Lippincott, Williams & Wilkins, Baltimore, Maryland, USA, 152 pp.

Shindel, N., Dardiri, A.H., Ferris, D.H., (1978). An indirect fluorescent antibody test for the detection of *Cytauxzoon*-like organisms in experimentally infected cats. *Canadian Journal of Comparative Medicine,* 42, 460-465.

Simking, P., Wongnakphet, S., Stich, R.W., Jittapalapong, S., (2010). Detection of *Babesia vogeli* in stray cats of metropolitan Bangkok, Thailand. *Veterinary Parasitology,* 173, 70-75.

Snider, T.A., Confer, A.W., Payton, M.E., (2010). Pulmonary histopathology of *Cytauxzoon felis* infections in the cat. *Veterinary Pathology,* 47, 698-702.

Soares, C.O., (2001). Cytauxzoonose felina e´ diagnosticada e isolada pela primeira vez na America Latina. *Revista Clınica Veterinaria*, 32, 56–58.

Soares, J.R., Izabel, S.A., Tercio, N.N., Scheide, R., Scofield, A., (2004). *Cytauxzoon felis*-like in the Mountain Lion (*Puma concolor*): A case report. *Journal of Animal and Vetrinary Advances,* 3, 820-823.

Solano-Gallego, L., Baneth, G. (2011). Babesiosis in dogs and cats - expanding parasitological and clinical spectra. *Veterinary Parasitology,* 181, 48-60.

Stewart, C.G., Hackett, K.J., Collett, M.G., (1980). An unidentified *Babesia* of the domestic cat (*Felis domesticus*). *Journal of the South African Veterinary Association,* 51, 219-221.

Sun, Y., Liu, G., Yang, L., Xu, R., Cao, W., (2008). *Babesia microti*-like rodent parasites isolated from *Ixodes persulcatus* (Acari: Ixodidae) in Heilongjiang Province, China. *Veterinary Parasitology,* 156, 333-339.

Susta, L., Torres-Velez, F., Zhang, J., Brown, C., (2009). An *in situ* hybridization and immunohistochemical study of cytauxzoonosis in domestic cats. *Veterinary Pathology,* 46, 1197-1204.

Tabar, M.D., Altet, L., Francino, O., Sánchez, A., Ferrer, L., Roura, X., (2008). Vector-borne infections in cats: molecular study in Barcelona area (Spain). *Veterinary Parasitology,* 151, 332-336.

Tsuji, M., Zamoto,A., Kawabuchi,T., Kataoka,T., Nakajima,R., Asakawa, M., Ishihara,C., (2006). *Babesia microti*-like parasites detected in Eurasian red squirrels (*Sciurus vulgaris orientis*) in Hokkaido, Japan. *Journal of Veterinary Medicine Science,* 68, 643-646.

Uilenberg, G., Franssen, F.F., Perié, N.M., (1987). Relationships between *Cytauxzoon felis* and African piroplasmids. *Veterinary Parasitology,* 26, 21-8.

Wagner, J. E., (1976). A fatal cytauxzoonosis-like disease in cats. Journal of the American Veterinary Medical Association, 168, 585–588.

Walker, D.B., Cowell, R.L., (1995). Survival of a domestic cat with naturally acquired cytauxzoonosis. *Journal of the American Veterinary Medical Association,* 206, 585 -588.

Wang, C., Ahluwalia, S., Li, Y., Gao, D., Poudel, A., Chowdury, E., Boudreaux, M., Kaltenboek, B., (2010). Frequency and therapy monitoring of canine *Babesia* spp infection by high resolution melting curve quantitative FRET-PCR. *Veterinary Parasitology*, 168, 11-18.

Weisman, J.L., Woldemeskel, M., Smith, K.D., Merrill, A., Miller, D., (2007). Blood smear from a pregnant cat that died shortly after partial abortion. *Veterinary Clinics Pathology,* 36, 209-211.

Weng, H.Y., Hart, L.A., (2012). Impact of the economic recession on companion animal relinquishment, adoption, and euthanasia: a Chicago animal shelter's experience. *Journal of Applied Animal Welfare Science,* 15, 80-90.

Widmer, C.E., (2009). Perfil sanitario des onças-pintadas (*Panthera onca*) da vida libre no Pantanal de Sul de Matto Grosso do Sul - Brasil. *PhD Dissertation,* College of Veterinary Medicine, University of Sao Paulo, (Sao Paulo, Brazil), pp. 1-78.

Yabsley, M.J., Murphy, S.M., Cunningham, M.W., (2006). Molecular detection and characterization of *Cytauxzoon felis* and a *Babesia* species in cougars from Florida. *The Journal of Wildlife Diseases*, 42, 366-374.

Yeagley, T.J., Reichard, M.V., Hempstead, J.E., Allen, K.E., Parsons, L.M., White, M.A., Little, S.E., Meinkoth, J.H., (2009). Detection of *Babesia gibsoni* and the canine small *Babesia* "Spanish Isolate" in blood samples from dogs confiscated from dogfighting operations. *Journal of the American Veterinary Association,* 235, 535-539.

Zahler, M., Rinder, H., Schein, E., Gothe, R., (2000). Detection of a new pathogenic *Babesia microti*-like species in dogs. *Veterinary Parasitology,* 89, 241-248.

Zamoto, A., Tsuji, M., Wei, Q., Cho, S., Shin, E., Kim, T., Leonova, G., Hagiwwara, K., Asakawa, H., Takashima, I., Ishihiara, C., (2004). Epizootiological survey for *Babesia microti* among small widl mammals in Northeastern Eurasia and a geographic diversitty in the beta-tubulin gene sequences. *Journal of Veterinary Medicine Science,* 66, 785-792.

Zobba, R., Parpaglia, M.L.P., Spezzigu, A., Pittau, M., Alberti, A., (2011). First molecular identification and phylogeny of a *Babesia* sp. from a symptomatic sow (*Sus scrofa* Linnaeus 1758). *Journal of Clinical Microbiology,* 49, 2321-2324.

Revised by Dr. Jeronimo Ruas, Laboratorio Regional de Diagnostico, Faculdade de Veterinária (LRD). Universidade Federal de Pelotas, P.O. Box 354 Pelotas, Brazil.

In: Felines
Editors: R. García and R. Jiménez
ISBN 978-1-62257-005-8

Chapter 3

EMERGING TICK-BORNE PROTOZOAL DISEASES - PART 2: FELINE HEPATOZOONOSIS AND THE "RESURRECTION" OF *HEPATOZOON FELIS*

Angel Criado-Fornelio
Parasitology Laboratory, Microbiology and Parasitology Department, Faculty of Pharmacy, Universidad de Alcalá, Alcalá de Henares, Spain

ABSTRACT

Hepatozoonosis is considered an emergent disease in felines. The genus *Hepatozoon* is a group of apicomplexan blood parasites found in a wide range of terrestrial vertebrates as well as many hematofagous invertebrates. Infection of the vertebrate host usually occurs through ingestion of an infected invertebrate host. Although hepatozoonosis in felines is normally subclinical, pathogenic effects may be exacerbated in stressed or immunocompromised animals. Concomintant infections may also worsen hepatozoonosis prognosis.

The advent of genetic identification procedures has extended the spectrum of known feline hepatozoons. Several methods, such as standard PCR, PCR-RFLP or real-time PCR are now available for the diagnosis of hepatozoonosis. A new method for diagnosis of *H. felis* and *H. canis* in cats is described in this chapter. This procedure is complementary to the quantitative PCR method described in the preceding chapter.

Hepatozoon sp from cats has been confounded for a long time with *H. canis*, due to its morphological similarity with the species present in dogs. Interestingly, DNA sequencing has demonstrated that infection caused by *Hepatozoon* spp (and also by *H. canis*) is relatively frequent in felines. There is controversy about the classification of some hepatozoons, due to the lack of differential morphologic features. In this respect, it is formally proposed in this chapter to confirm the denomination of *Hepatozoon felis* (Patton, 1908) for the two isolates of domestic cats previously sequenced in our laboratory (Criado-Fornelio *et al.*, 2006) and therefore, for those isolates genetically related to them. Such proposal is based on the different host range shown by *H. felis* and *H. canis*. This biological trait was confirmed by contingency table analysis. The idea is in agreement with the principle of priority in Zoological Taxonomy and moreover, is supported by the application of the molecular clock hypothesis to estimate divergence dates between species. In addition it is hypothesized that *Riphicephalus sanguineus* is one of the most likely vectors for *H. felis*, based on their common geographic distribution and their similar frequency in felines.

Although arthropods are the suspected definitive hosts for all feline hepatozoons, the application of vector control measures may have a limited efficacy for reducing disease levels in domestic cats and may prove very difficult to apply for wild felines. Due to the fact that hepatoozoonosis in cats is usually subclinical, there is no standard chemotherapeutic protocol available. There are, however, some reports of successful treatments with antibiotics and primaquine.

1. INTRODUCTION

Hepatozoons infect a wide variety of vertebrates. Species of *Hepatozoon* are common in marsupials, insectivores, leporids and very common in rodents and carnivores. In fact, two species infecting dogs and cats (*Hepatozoon canis* and *H. felis*) have been recorded in domestic animals and wildlife in many parts of the world (Smith, 1996). The life cycle of the hepatozoons is indirect. Vertebrates are intermediate hosts and insects or acarines serve as definitive hosts. *Hepatozoon* species show in general low host specificity, with some exceptions (Telford *et al.*, 2001).

The first description of *Hepatozoon* in felines dates back to 1908, when Patton described the species *Hepatozoon felis* in the domestic cat. Reports of hepatozoonosis in felines are not as common as in canids (Mercer *et al.*, 1988,

Baneth *et al.*, 1998a, Criado Fornelio *et al.*, 2006, Jittapalapong *et al.*, 2006). This may be due to the fact that hepatozoonosis is often a subclinical infection in cats and their kin. Disease presentation in cats may be summarized as dehydration, weight loss, lethargy, ocular discharge, anorexia and icterus. Since species of *Hepatozoon* are opportunistic pathogens, felines under stress or co-infected with other pathogens may show signs of disease.

Hepatozoonosis has been traditionally diagnosed by microscopy. Some specialized laboratories have developed immunological procedures for antibody detection in dogs (Gonen *et al.*, 2004), but these are not comercially available and moreover, they have not been validated for use in felines. The same is true for molecular procedures. However, the latter have contributed to widen our knowledge on hepatozoonosis epizootiology. It is now well-established that carrier animals are relatively frequent in feline populations. This is more evident in wild felids or stray cats (Ortuño *et al.*, 2008). In addition, genetic characterization of isolates (based on DNA sequencing) has led to the discovery of some new hepatozoons of felines. Their taxonomic status is still controversial, since data on their host range (both intermediate and definitive) and other differential traits are not yet available. Some authors have suggested that genetic divergence seems to point out possible new species (André *et al.*, 2010). In contrast some researchers have suggested that more data on epizootiology (both on the definitive and intermediate host) are needed before drawing any definitive conclusions on the taxonomic status of the recently discovered hepatozoons (Baneth, 2011).

In the present chapter, we shall review different aspects of hepatozoonosis in felines, such as the morphology, biology, host diversity, pathology, epidemiology and diagnosis, taxonomy, phylogeny and preventive/control measures.

2. Feline Hepatozoons: Morphology and Life Cycle

The available information on the morphology and biology of feline hepatozoons is scarce. The gamont stage found in cat blood is similar to that of *H. canis* (Baneth *et al.*, 1998a), and molecular studies indicate that indeed this species is present in felines. Meronts (schizonts) of *Hepatozoon* sp have been observed in feline heart tissue (Kubo *et al.*, 2006). These are morphologically similar to those found in dog tissues (bone marrow), as described by Baneth *et al.* (2007). The latter authors described as well the stages of *H. canis* in *Riphicephalus sanguineus*. However no studies on the morphology of other mammalian *Hepatoozon* spp in the arthropod vector have been published up to now.

The life cycle of feline hepatozoons has not been studied in detail. For *H. canis*, it is assumed that it might be similar to that occurring in dogs. Hence, an indirect life cycle where cats are the intermediate hosts and ticks or fleas are the definitive hosts is likely to occur in Nature. Sexual reproduction (final gamete formation plus fertilization with subsequent schyzogony) is linked to the arthropod host, whereas schizogony (merogony) and initial gametogenesis occurs in the mammalian host. Based on the life cycle of *H. canis* in dogs described by Baneth *et al.*, (2007), along with data published by Kubo *et al.*, (2006), a draft of the parasite' life cycle in cats is presented in Figure 1. In summary, the life cycle involves felines (as intermediate hosts) and the brown dog tick, *Rhipicephalus sanguineus* (as definitive host). Although the common name of the tick suggests that cats are not suitable hosts for these acarines, *R. sanguineus* may be collected both from cats and dogs (Ortuño *et al.*, 2008; Berrada and Telford, 2009; Mendes *et al.*, 2011). There have been reports of diverse *Hepatozoon* sp infections in felines, but there is no information available on their possible definitive hosts; indeed ticks, fleas or other artrhopods are good candidates (Criado-Fornelio *et al.*, 2006; Salakij *et al.*, 2008 and 2010; Bortoli *et al.*, 2011).

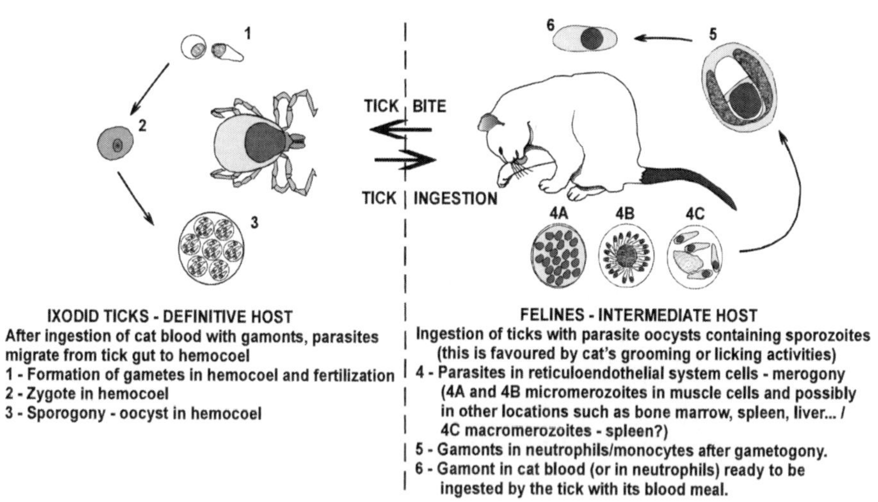

Figure 1. Schematic representation of the life cycle of the feline hepatozoons.

The life-cycle of feline hepatozoons is completed by ingestion of an infected tick by the definitive host. The ingestion of ticks may occur in several ways, such as grooming, licking, scavenging or by predation. But tick ingestion is not the only way to complete the life cycle of hepatozoons (Figure 2). There are two alternative ways of transmission of *H. canis*, although they have been demonstrated only in dogs. It seems that some *Hepatozoon* stages living in host tissues (cystozoites) may survive after ingestion and subsequently boost an infection in the new host, thus helping to spread the disease (McCully *et al.*, 1975; Smith, 1996; Baneth and Shkap, 2003). In addition, vertical transmission of *H. canis* has been demonstrated in puppies by Murata *et al.*, (1993).

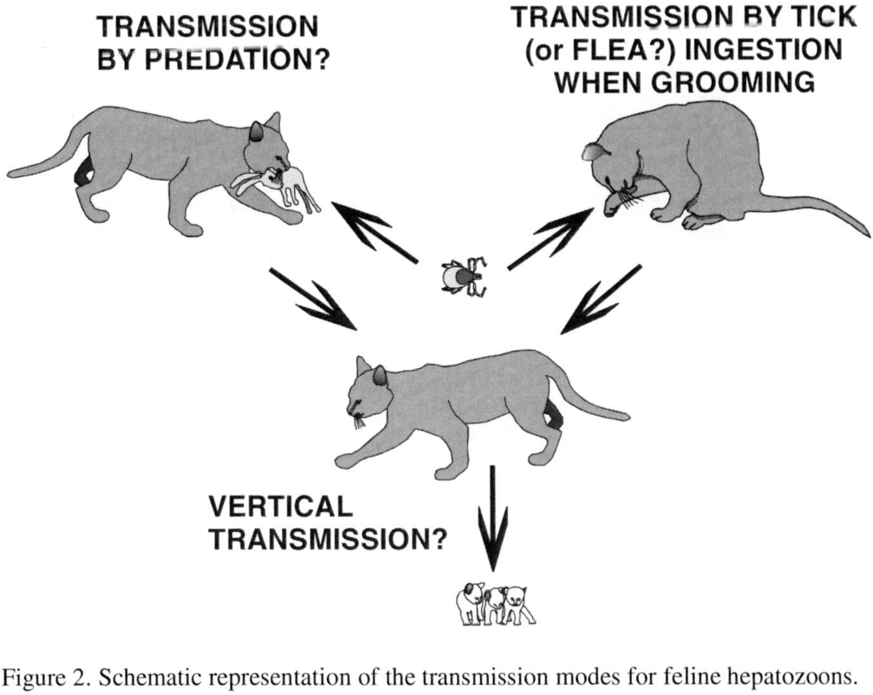

Figure 2. Schematic representation of the transmission modes for feline hepatozoons.

3. Feline Hepatozoons: Pathology

Feline hepatozoonosis is usually subclinical. Most studies conducted on cats have found that infected animals were asymptomatic. Former publications by Klopfer *et al.*, (1973), Ewing (1977), Van Amstel (1979) or Baneth *et al.*, (1995) were focused on individual cases. In a more detailed study, Baneth et al

(1998a) described the clinical symptoms observed in a group of diseased animals (7 cats). A list of publications where clinical signs were reported in *Hepatozoon*-infected felines is presented in Table 1.

Taking into account these contributions, it is evident that the classic clinical presentation of feline hepatozoonosis consist of fever (or hypothermia), renal failure, dehydration, respiratory problems, enlarged lymph nodes, weight loss, lethargy and ocular discharge.

In *Hepatozoon*-infected cats, some blood biochemistry parameters may also show noticeable alteration. It is common the apparition of increased levels of creatine phosphokinase, lactate dehydrogenase, and sometimes of alanine aminotransferase (Baneth *et al.*, 1998a). However, the clinical, biochemical and hematological findings detected in feline hepatozoonosis may be highly variable. This is more evident if there are concomitant infections casued by feline retroviral infections (such as Feline Immunodeficiency Virus – FIV or Feline Leukemia Virus – FeLV) and *Mycoplasma* spp (Baneth, 2003; Ortuño *et al.*, 2008).

Table 1. Summary of clinical findings related to feline hepatozoons in recent publications

Publication	Main clinical signs	Host / *Hepatozoon* species diagnosed
Baneth *et al.*, (1998a)	Anorexia, lymphadenitis, dehydration, renal and respiratory signs, ataxia. Increased LDH and CK. Concomitant viral infections	Domestic cat / *Hepatozoon sp* (diagnosed by microscopy)
Kubo *et al.*, (2006)	Histopathological alterations in the hearth	*Felis triomotensis* / *Hepatozoon* sp (diagnosed by microscopy)
Ortuño *et al.*, (2008)	Respiratory signs. Concomitant FIV infection	Domestic cat / *Hepatozoon felis* (sequencing)
Tabar *et al.*, (2008)	Clinical symptoms (Not specified by the authors)	Domestic cat / *Hepatozoon felis* (sequencing)
East *et al.*, (2008)	Ataxia, labored breathing	*Crocuta crocuta* (Feliformia) / *Hepatozoon felis* (sequencing)

It has been pointed out that stressful conditions may reactivate *Hepatozoon* infection. Such possibility is extremely important for wild feline populations (Baneth *et al*, 1998a; Beaufils, 2004; Harrus and Baneth, 2005; Kubo *et al*., 2006 and Metzger *et al*, 2008).

4. NEW ADVANCES IN THE DIAGNOSIS OF FELINE HEPATOZOONS

The traditional diagnosis method for feline hepatozooonosis is microscopy. As stated before, *Hepatozoon* gamonts may be observed in feline blood, but unfortunately, they are found inside erythrocytes in limited numbers. Approximately 1-2% of peripheral blood neutrophils are infected with gamonts, according to Baneth *et al*., (1998a) and Salakij *et al*., (2008). In wild felines, it seems that the percentage of infected neutrophils is even lower (Mercer *et al*., 1988). Hence, microscopic observation usually fails to detect the parasite (Kubo *et al*., 2006).

At this moment there are no immunological procedures commercially available for the detection of *Hepatozoon* sp in felines. In contrast, some attempts have been published for diagnosis of canine hepatozoonosis, albeit all of them are still experimental. Baneth *et al*., (1998b) tested the IFAT technique for diagnosis of *H. canis* in dogs. Matthew *et al*., (2001) developed an ELISA test for the detection of *H. americanum* in dogs. Gonen *et al*., (2004) employed another ELISA test for detection of *H. canis* in dogs and the same research group has tested this procedure for detection of the same parasite in red foxes (Fishman *et al*., 2004).

A certain number of molecular methods have been developed for the detection of hepatoozons. Although these are not specifically designed for felines, they might likewise work for them. Before summarizing the published procedures, it is important to underline that Li *et al*., (2008) found (in their investigations on the set up of a qPCR assay for diagnosis of *H. canis* in dogs) that diagnostic sensitivity increased by 300% when blood samples were maintained at 37 ºC for three days. The reason for this effect is unclear, but it is a wise idea for those willing to diagnose *Hepatozoon* sp in felids to use such incubation conditions for their blood samples before proceeding with DNA extraction.

A summary of the main molecular procedures available for diagnosis of *Hepatooozon* spp is presented in Table 2.

Table 2. Summary of molecular methods of diagnosis for feline/canine hepatozoons

Publication	Procedure	Host(s) / *Hepatozoon* species diagnosed
Inokuma *et al.*, 2002	Partial amplification of the 18S rRNA gene - Standard PCR	Dog / *H. canis*
Criado-Fornelio *et al.*, (2003b)	Partial amplification of the 18S rRNA gene - Standard PCR	Dog and fox / *H. canis*
Criado-Fornelio *et al.*, (2006)	Complete amplification 18S rRNA gene – Standard PCR	Canids and cats / *H. canis* and *H. felis*
Criado-Fornelio *et al.*, (2007a)	PCR-RFLP Partial amplification 18S rRNA gene	Dog / *H. canis*
Criado-Fornelio *et al.*, (2007b) and (2009)	Partial amplification 18S rRNA gene – Real time PCR with SYBR green	Dog and cat / *H. canis* and *Hepatozoon* sp
Li *et al.*, (2008)	Partial amplification 18S rRNA gene – Real time PCR with FRET probes	Dog / *H. canis* and *H. americanum*
Allen *et al.*, (2008)	Partial amplification 18S rRNA gene – Standard PCR (Test not specific for *Hepatozoon*)	Dog / *Hepatozoon* spp
Kledmanee *et al.*, (2009)	Partial amplification 18S rRNA gene – Multiplex PCR	Dog / *H. canis* spp + *Babesia* spp + *Ehrlichia canis*

The available diagnostic assays for hepatozoons are based on three methods:

1) Standard PCR with detection of amplified products by DNA electrophoresis in agarose gels. Sometimes this method has been combined with restriction enzyme digestion (PCR-RFLP, as reported by Criado-Fornelio *et al.*, 2007a). In addition, a multiplexed DNA detection assay (primarily intended for diagnosis of *H. canis*, *Babesia* spp and *Ehrlichia canis* in dogs) has been published by Kledmanee *et al.*,(2009).

```
                                 ---------Sensor probe--------      --------Anchor probe-------- / No. differences
AY628681 H. felis Spain 2        CCATGCACGTCTAAGTATAAACAGTAATA-CAGT-AAAACTGCAAATGGCTCATTAAAACAGT   0 = PROBE
AY620232 H. felis Spain 1        CCAGGCACGTCTAAGTATAAACAGTAATA-CAGT-AAAACTGCAAATGGCTCATTAAAACAGT   1
AY461378 H. canis Spain 2        CCATGCACGTCTAAGTATAAACAAAAATA-CAGT-AAAACTGCAAATGGCTCATTACAACAGT   3
AF176836 H. americanum           CCATGCATGTCTAAGTACATACAATAATA-CAGT-AAAACTGCAAATGGCTCATTAAAACAGT   4(not tested)
AY309956 Cytauxzoon sp           CCATGCATGTCTAAGTATAAGCTTTTATA-TGGT-GAAACTGCGAATGGCTCATTAAAACAGT   6
AY150068 Theileria annae         CCATGCATGTCTTAGTATAAGCTTTTATA-CAGC-GAAACTGCGAATGGCTCATTAAAACAGT   8
AY150057 B. canis canis          CCATGCATGTCTAAGTACAAGCTTTTTA -CGGT-GAAACTGCGAATGGCTCATTACAACAGT  11
                                 *** *** **** **** ** *               ******* *********** ******
```

Figure 3. Alignment of *Hepatozoon* spp 18S rRNA gene sequences targeted by the FRET probes. Only the hepatozoons and piroplasmids most commonly found in European felines were included in the alignment. The GenBank® accession numbers for piroplasmid sequences, FRET probe binding regions (both anchor and sensor), and the number of base differences (compared to FRET probes) are indicated. Conserved positions in sequences are shown by asterisks. Variable positions in sequences are underlined. *H americanum* is included in the alignment but was not tested in diagnostic assays.

2) Real-time PCR (qPCR) with SYBR green.
3) Real-time PCR with oligonucleotidic probes (assayed in dogs only).

Standard PCR methods shall not be discussed here in detail. With regard to qPCR assays, the SYBR green-based procedure described by Criado-Fornelio *et al.*, (2007b) has the advantage of avoiding electrophoretic analysis by using a cheap intercalating agent instead of expensive oligonucleotidic probes. For those willing to diagnose felids in the American continent, the test of Li *et al.*, (2008) has the advantage of species recognition by FRET probes, although it has not been tested in felines. Therefore, additional data on its sensitivity and specificity are needed to confirm its performance.

A new molecular method for multiple detection of feline hemoprotozoa (piroplasmids and hepatozoons) was presented in the chapter "Emerging Protozoal Tick-Borne Diseases - Part 1" (in this book). In addition a new qPCR procedure for unequivocal diagnosis of *H. felis* and *H. canis* in felines is described in this chapter. The method is quite similar to the procedure described by Criado-Fornelio *et al.*, (2009b). Based on the sequence of *H. felis* (Spain-2), the primers UPFR-F = 5'-ttgatcctgccagtagt-3' and UPFR-R=5'-agctgatrggtcagaaac-3 were combined with FRET probes HEF (5'-actgttgtaatgagccatttgcagtttt-3' Fluorescein) and HE705 (5'LCRED 705-tattactgtttatacttagacgtgcatgg-3'). The FastStart DNA MasterPlus Hybprobe Kit (Roche, Mannheim, Germany) was used in qPCR assays following the manufacturer's instructions. Assay's specificity was tested as previously reported (Criado-Fornelio *et al.*, 2009b), by using DNA samples of *Toxoplasma*, *Neospora* and *Leishmania*-infected animals as negative amplification controls. Positive amplification controls were DNA samples of carnivores infected by *Cytazuxzoon* sp, *Theileria annae*, *B. canis canis*, *B. canis vogeli*, *Hepatozoon felis* and *H. canis*), which had been diagnosed as such in previous studies (Criado-Fornelio *et al.*, 2003a, 2003b, 2004 and 2006).

The optimal thermocycling conditions were identical to those described in the qPCR method in the preceding chapter ("Emerging Protozoal Tick-Borne Diseases-Part 1" in this book). Serial dilutions of linearized p-GEM plasmid with *H. felis* Spain-2 DNA inserts (corresponding to the target region) were used in these performance assays. The proposed qPCR assay showed good specificity, since no amplification products were detected in negative control samples, whereas strong amplification signal was noticed with all *Hepatozoon*-positive samples. In contrast, amplification curves for piroplasmids showed weaker amplification signal due to the six or more mismatches of the fluorescence probe (the sequence alignment is shown in Figure 3).

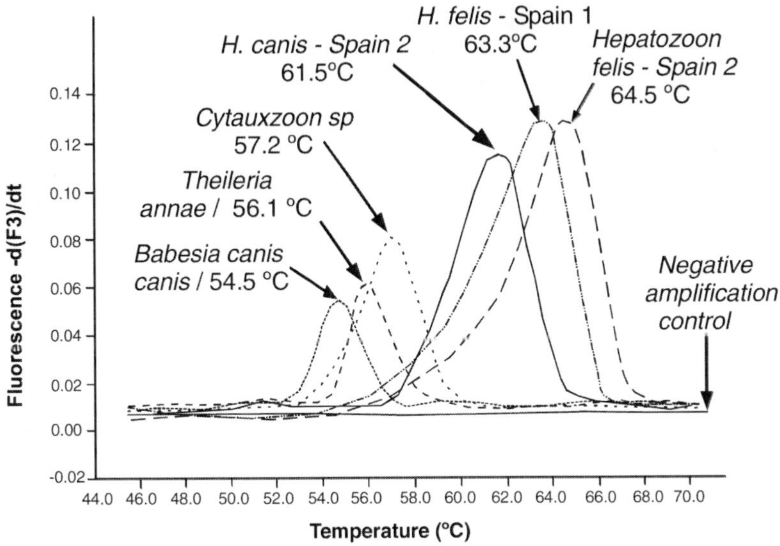

Figure 4. Melting curve analysis obtained in the FRET probe method. Curves obtained with control samples (*Hepatozoon canis* (Spain 2), *H. felis* (genotypes Spain 1 and 2) *Babesia canis canis*, *B. canis vogeli*, *Cytauxzoon* sp, *Babesia* (*Theileria*) *annae*), and negative control (*Leishmania infantum*).

The Student t test was used to confirm the existence of statistically significant differences between melting peak temperatures of *Hepatozoon*/ piroplasmid species. The significance in statistical analysis was set at P<0.001 or lower. Due to sequence differences in the target region of the probe, the FRET assay readily distinguished between *H. felis* and *H. canis*, whereas piroplasmid species showed lower melting temperatures (Figure 4). However, discrimination of feline piroplasmid species was not an objective of this work, since a different assay had been previously set up for such purpose. The new qPCR method was not tested in *H. americanum* since no samples of that species were available in our laboratory. Theoretically (Figure 3) the assay should be able to discriminate *H. canis*, *H. felis* and *H. americanum*. However, a new primer (UPFR2-F 5'- cagtagtcatatgcttgtc-3', which is a trimmed version of the primer HAM-1(f) of Criado-Fornelio *et al.*, 2006) should be included in the reaction mix in order to obtain amplificacion in *H. americanum*. This is due to sequence differences between *H. canis*/ *H. felis* and *H. americanum*.

The sensitivity of the proposed technique was about 20 copies of plasmid DNA. The assay provided linear response curves over the $10^3 - 10^8$ copies range. The standard curve, determined with the Lightcycler software, showed excelent performance (correlation coefficient = 0.998 and slope -3.21); these data suggested that the qPCR assay had been set up correctly.

The new assay was used for screening 450 blood samples of domestic cats from Spain. The samples were obtained by veterinarians in a local animal shelter and then send for diagnosis to our laboratory. Extraction of DNA from mammalian blood was performed with the MOBIO Blood Spin DNA kit (Mobio, Solana Beach, CA, USA). The animals studied were either asymptomatic or showed some signs of disease, usually not related to piroplasmosis. The qPCR technique showed that 3 cats were positive for *H. felis*, whereas standard PCR (Criado-Fornelio *et al.*, 2007a) only detected one infected cat. In addition, the alternative qPCR assay (as described in the chapter "Emerging Protozoal Tick-Borne Diseases - Part 1" - in this book) detected 2 of the *Hepatozoon*-infected cats, whilst the amplification curve obtained for the remaining infected animal was apparently negative. Nonetheless, a very small peak corresponding to *Hepatozoon* sp was observed in graphics after performing the corresponding melting curve analysis. DNA sequencing showed that all isolates were genetically identical to those previously described in cats from Spain. Therefore, the qPCR assay described in this chapter is more sensitive and versatile than that described in the preceding chapter for diagnosis of feline hepatozoons. Moreover it might be comparable to the assay described by Li *et al.*, (2008) for detection of *H. canis* and *H. americanum*, in dogs. Nonetheless, these authors did not study the performance of the test for diagnosis of *H. felis*.

5. Epizootiology of Feline Hepatozoonosis

Before the advent of molecular methods of diagnosis, the amount of information available on the epizootiology of feline hepatozoons is quite limited. The parasite, identified as *Hepatozoon* sp or *H. canis* was reported in countries such as South Africa (Van Amstel, 1979), the USA (Metzger *et al.*, 1988) and Israel (Baneth *et al.*, 1995). Since PCR plus sequencing techniques detected *Hepatozoon*-infected cats for the first time (Rubini *et al*, 2006; Jittapalapong *et al.*, 2006; Criado-Fornelio *et al.*, 2006), many reports confirmed the presence of this hematozoan in felines. Unfortunately, there is some degree of uncertainity in some of the molecular identifications performed. This is due to the fact that the sequence information obtained was limited to the central part of the 18S rRNA gene. The 5' end of the gene is considered the most suitable region for identification purposes ("DNA barcoding") in many organisms (Bhadury *et al.*, 2006). A summary of publications on the epizootiology of feline hepatozoons is shown in Table 3. It is interesting to underline that sometimes *Hepatozoon* species typical of frogs or snakes have been detected in felines (Salakij *et al*, 2008 and 2010). Moreover, it has been pointed out that some new feline *Hepatozoon* species might be present in the American continent and India (André *et al.*, 2010, Allen *et al.*, 2011; Pawar *et al.*, 2012).

Table 3. Summary of epizootiologic findings related to feline hepatozoonosis. Emphasis is given to recent publication based on molecular methods

Publication	Country	Host(s) / *Hepatozoon* species diagnosed	Diagnostic method / Prevalence (No. animals studied)
Van Amstel (1979)	South Africa	Domestic cat / *Hepatozoon* sp	MIC / NA (1)
Mercer *et al.*, (1988)	USA	Domestic cat / *Hepatozoon* sp	MIC / NA (1)
Averbeck *et al.*, (1990)	Tanzania	*Panthera leo* and *Acinonyx jubatus* / *Hepatozoon* sp	MIC / 100% and 100% (123 and 8)
Baneth *et al.*, (1998b)	Israel	Domestic cat / *Hepatozoon* sp	MIC / 0.57 % (1229)
Kubo *et al.*, (2006)	Japan	*Felis triomotensis* and *F. bengalensis* / *Hepatozoon* sp	MIC / 56 % (30) and 14 % (42)
Jittapapalong *et al.*, (2006)	Thailand	Domestic cat / *H. canis*	PCR+S / 0.7 % (300)
Rubini *et al.*, (2006)	Brazil	Domestic cat / *H. canis*	PCR+S / NA (3)
Criado-Fornelio *et al.*, (2006)	Spain	Domestic cat / *H. felis*	PCR+S / 0.6 % (300)

Table 3. (Continued)

Publication	Country	Host(s) / *Hepatozoon* species diagnosed	Diagnostic method / Prevalence (No. animals studied)
Ortuño *et al.*, (2008)	Spain	Domestic (stray) cat / *H. felis*	PCR+S / 16 % (25)
Tabar *et al.*, (2008)	Spain	Domestic cat / *H. felis*	PCR+S / 4 % (100)
Metzger *et al.*, (2008)	Brazil	*Leopardus* spp - *Puma yaguaroundi* / *H. felis*	PCR+S / 3.4 % (29)
Salakij *et al.*, (2008)	Thailand	*Prionalurus planiceps* / *H. catesbianae*	PCR+S / NA (1)
Criado-Fornelio *et al.*, (2009a)	Venezuela	*Panthera onca* / *H. canis*	PCR+S / NA (2)
Criado-Fornelio *et al.*, (2009b)	France	Domestic cat / *H. canis*	PCR+S / 1.7 % (116)
Widmer (2009)	Brazil	*Panthera onca* and diverse ticks (*Amblyomma cajennense*, *A. triste* and *Riphicephalus microplus*) / *H. felis*	PCR+S / 100% (10) and 100% (40)
Salakij *et al.*, (2010)	Thailand	*Prionalurus bengalensis* / *Hepatozoon sp*(python)	PCR+S / NA (2)
André *et al.*, (2010)	Brazil	*Leopardus tigrinus*, *Puma* spp / *Hepatozoon sp*	PCR+S / 6 % (100)
Kubo *et al.*, (2010)	Korea	*Prionalurus bengalensis* / *H. felis*	PCR+S / NA (5)
Sakuma *et al.*, (2011)	Japan	*Prionalurus bengalensis* / *H. felis*	PCR+S / NA (1)
Allen *et al.*, (2011)	USA	*Felix rufus* / *H. felis* and domestic cat / *H. americanum*	PCR+S / NA (3)
Bortoli *et al.*, (2011)	Brazil	Domestic cat / *H. felis*	PCR+S / 0.5 % (200)
Pawar *et al.*, (2012	India	Domestic cat and diverse wild felids / *H. felis*	PCR+S / 50% (22)

Abbreviations: NA = not applicable (case report, limited number of animals studied and prevalence not feasible, etc); MIC = microscopy; PCR+S = PCR and sequencing.

It is not clear if the prevalence of *H. canis* in dog correlates with the infection rate in cat. In Spain, it seems that in regions where *H. canis* appears in dog, the parasite is absent in cat, whilst the latter may be infected by *H. felis* instead (Criado-Fornelio, unpublished data). Prevalence of feline hepatozoonosis is variable depending on parasite, host and location (Table 3). In temperate countries, low prevalence of *H. felis* is usually reported in domestic cats (0.6 to 4%, according to Criado-Fornelio *et al.*, 2006 and Tabar *et al.*, 2008). However, higher infection rate (16%) has been found in stray cats in the same region (Ortuño *et al.*, 2008). In tropical countries there are evident differences: *H. felis* may be absent (Carneiro, 2007) or present with low or moderate prevalence (Bortoli *et al.*, 2011; Pawar *et al.*, 2012). In contrast, *H. canis* show low prevalence both in temperate (1.7% in France, according to Criado-Fornelio *et al.*, 2009b) and tropical countries (0.7% in Thailand, according to Jittapalapong *et al.*, 2006). This suggests that in spite of the low parasitic specificity shown by hepatozoons (Smith, 1996), there is some kind of host preference, as it will be discussed in the following section. In general, studies in wildlife show high disease rates (from 3.4 to 100%, as reported by Metzger *et al.*, 2008 and Widmer, 2009). This may be caused by frequent exposure to ticks in the wild, which facilitates ingestion when grooming or by predation.

Little research efforts have been devoted to determine the arthropod vectors of feline hepatozoons. It is suspected that *Riphicephalus sanguineus* is the definitive host for *H. canis*, and moreover, indirect evidence indicates that it might be as well for *H. felis*. If we consider the surveys where presence of ticks and feline hepatozoons were studied, it is evident a noticeable correlation between prevalence of *Rhiphicephalus spp* and *H. felis*:

a) In Spain, Ortuño *et al.*, (2008): *Riphicephalus* spp = 28% (*R. sanguineus* = 20%) - *H felis* (in cat) = 16%;
b) In Brazil:
   - Carneiro (2007): no ticks – no infected cats;
   - Widmer *et al.*, (2009): *R. sanguineus* = 100% - *H. felis* (in jaguar) = 100 %;
   - Bortoli *et al.*, (2011): *R. sanguineus* = 0.5% – *H. felis* (in cat) = 0.5%.

Additional evidence suggests that *H. felis* is transmitted by *Riphicephalus* sp. A glance on the geographic distribution of feline hepatozoons (Figure 5) shows that *H. felis* and *H. canis* are found worldwide. The same is true for *B.*

canis vogeli (Solano-Gallego and Baneth, 2011). Since *H. canis* and *B. canis vogeli* are transmitted by *R. sanguineus* (Baneth, 2011; Solano-Gallego and Baneth, 2011), it is logical to infer that *H. felis* might be transmitted by the same arthropod vector. The cosmopolitan distribution of *Riphicephalus* species has already been pointed out in the literature (Hopla *et al.*, 1994). Further research is granted to address such hypothesis.

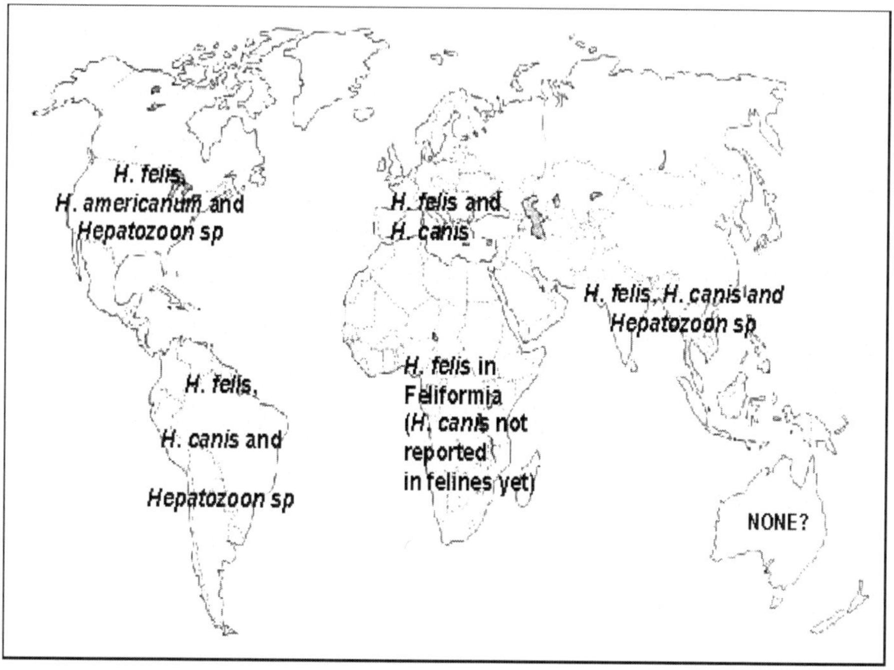

Figure 5. Approximate geographic distribution (by continent) of the feline hepatozoons.

Studies in wild felines from Brazil and India (André *et al.*, 2011; Pawar *et al.*, 2012) reported no ticks on the animals surveyed, whereas prevalence of hepatozoonosis was low to moderate (3.6 and 50%, respectively). The authors speculated about the possibility that these protozoa may have been transmitted by the ingestion of infected prey, as also suggested by East *et al.*, (2008) and Metzger *et al.*, (2008). It is evident that the relative importance of different ways of transmission of hepatozoonosis will remain as mere speculation until further investigations on the arthropods vectors of the feline hepatozoons are conducted.

6. Notes on Taxonomy and Phylogeny of *Hepatozoon* Species of Felines: "The Resurrection" of *Hepatozoon Felis*

As stated in the introduction section of this chapter, the first *Hepatozoon* species described in domestic cats is *H. felis* (Patton, 1908; Smith, 1996). Due to the lack of genetic analysis at that time, Wenyon (1926) suggested that it was impossible to distinguish between the hepatozoons found in cats and those found in canids. Therefore, he concluded that *H. canis* might be the best denomination applicable to feline hepatozoons. The matter remained unchanged for many years, although some doubts persisted. In fact, Smith (1996) in his review of the genus *Hepatozoon* indicated that three *Hepatozoon* species (namely *H. muris*, *H. canis* and *H. felis*) had been diagnosed in domestic and laboratory animals worldwide. This supports the idea that the old name of *H. felis* was never totally rejected. Bowman *et al.*, (2001) indicated that *H. felis* is the species present in domestic cats. However, they pointed out that for some authors this specific name was a synonym of *H. canis*, and moreover, they indicated that feline hepatozoons may correspond rather to *H. canis* or *H. procyonis*. The state of confusion remained unchanged until Criado-Fornelio *et al.*, (2006) sequenced for the first time the 18S rRNA gene of two *Hepatozoon* sp isolates from Spanish cats. Both Spanish isolates showed 96% sequence identity with *H. canis*, and therefore they were introduced in the GenBank® database with the name of *H. felis*, since it was evident that they possibly belonged to a different species (at least from the genetic point of view). However, the lack of additional information prompted its provisional identification in the original publication as *Hepatozoon* sp. Some time later, East *et al.*, (2008) found another *Hepatozoon* isolate (closely related to the Spanish *H. felis*) in spotted hyenas (*Crocuta crocuta*) from Tanzania. The African isolate was morphologically identical to *H. canis* of dogs. In South America, Metzger *et al.*, (2008) reported the presence of a *Hepatozoon* sp gamont in a blood smear form a Brazilian ocelot and it showed lesser size than that of *H. canis* found in domestic cat in Brazil, as reported by Perez *et al.*, (2004). However, the short fragment sequenced by Metzger *et al.*, (2008) makes uncertain the identification of this feline isolate. Perez *et al.*, (2004) reported the presence of *Hepatozoon* sp in Brazilian cats, but the isolate was not studied at genetic level. Since most feline hepatozoons seem to be morphologically very similar, the only way to obtain accurate identification relies on molecular methods.

In this chapter it is formally proposed that the name of *Hepatozoon felis* (Patton, 1908) should be used to define organisms showing high degree of genetic similarity with *Hepatozoon* sp from Spain (GenBank® accession numbers AY620232 and AY628681, identified already as *H. felis*).

This proposal is based on the following evidence:

1) Current data indicates that *H. felis* is very common in felids and very rare in canids. To provide statistical support to the hypothesis of relative host specificity, a contingency table (Table 4) was prepared by compiling data from diverse publications dealing with *Hepatozoon* infection in carnivores, namely the following authors: Oyamada *et al.*, (2005); Jittapalapong *et al.*, (2006); Rubini *et al.*, (2006); Criado-Fornelio *et al.*, (2006); Ortuño *et al.*, (2008); Tabar *et al.*, (2008); Metzger *et al.*, (2008); Criado-Fornelio *et al.*, (2009b); Spolidorio *et al.*, (2009); Tabar *et al.*, (2009), Götsch *et al.*, (2009); Vojta *et al.*, (2009); Widmer, (2009); André *et al.*, (2010); Allen *et al.*, (2011); Bortoli *et al.*, (2011), Abd-Rani *et al.*, (2011), Vargas-Hernández, *et al.*, (2011). The only canid (a South American fox) found infected by *H. felis* has been reported by Giannitti *et al.*, (2012). It is important to underline that the presence of *H. felis* in that fox was possibly favoured by a concomitant canine distemper virus infection that weakened the animal. In total, these publications have studied 1282 felines (of which 136 were infected) and 2657 canids (of which 581 were infected). Although *H. felis* has been found in spotted hyenas (East *et al.*, 2008), it is important to remind that this species is more closely related to felids (suborder Feliformia) than to canids (Suborder Caniformia), according to Eizirik *et al.*, (2010). To avoid any controversy, the report of the infection diagnosed in hyaenas was omitted from the analysis. Contingency analysis demonstrated (with a pretty high statistical significance –Table 4) that *H. canis* is a typical parasite of canids while *H. felis* is a typical parasite of felines. Such partial host specificity is supported by studies conducted in Europe and Asia. In Spain more than 500 dogs and 900 cats have been studied by several authors (Criado-Fornelio *et al.*, 2006; Ortuño *et al.*, 2008; Tabar *et al.*, 2008, Criado-Fornelio and 2009b and Criado-Fornelio, data reported in this chapter and unpublished data) and it was found that all cats were infected by *H. felis* whereas all dogs were infected by *H. canis*. The same was true in a recent study conducted in Indian wild and domestic felids and canids (Pawar *et al.*, 2012). This

suggests the existence of relative host specificity for *H. felis*, as it has been pointed out as well for some other hepatozoons (Telford *et al.*, 2001).

Table 4. Contingency table analysis of parasitization by *H. canis* and *H. felis* in infected canids (mainly dogs plus one South American fox; n = 581) and infected felines (both domestic and wild animals; n = 136)

Parasite Species	No. infected hosts (FELINES)	No. infected hosts (CANIDS)	Total
H. felis	34	1	35
H. canis	102	580	682
TOTAL	136	581	717

Note: Yates correction was applied. The significant association - *H. canis* in dogs and *H. felis* in felines - was confirmed by the Pearson ($P < 0.00001$) and χ^2 ($P < 0.00001$) tests. Data obtained from diverse publications (see chapter text) were analyzed with the STATGRAHICS® software package. The total number of animals studied by various authors was 2657 canids and 1282 felids (including infected and uninfected individuals).

2) The sequence of the 18S rRNA gene of *H. canis* and *H. felis* showed a genetic divergence of 4 % (Criado-Fornelio *et al.*, 2006), which is similar (or even greater) than those found in other cases where new species of parasitic protozooa of felines have been proposed and accepted. There are three clear examples:
   a) *Cytauxzoon manul* was proposed as a new species, even though there were no morphological differences with rapport to *C. felis* (Reichard *et al.*, 2005). However, there was 1.49 % corrected sequence divergence between the 18S rRNA gene sequences of *C. manul* and *C. felis*. Such feature, together with the fact that the parasite was found in a new feline host (*Otocolobus manul*) was considered a good base for the erection a new species. Interestingly, some time afterwards *C. manul* was found to be capable of infecting domestic cats as well (Joyner *et al.*, 2007).
   b) Another species proposed in cats is *Babesia canis presentii* (Baneth *et al.*, 2004). In this case there were some evident morphological differences between the new species and *B. canis canis*. But the genetic divergence in the 18S rRNA gene reaches only 1 % when analyzed by BLASTN® (BLAST two sequences).

However, a pairwise sequence analysis conducted by these authors showed 3% divergence. Studies on the ITS sequences further demonstrated the existence of evident genetic divergence, which provided support to the existence of a new subspecies.

c) Finally, *Babesia lengau* from cheetah (*Acinonyx jubatus*) showed 4% divergence (in the sequence of the 18S rRNA gene) with regard to *B. conradae/B. duncani* (Bosman *et al.*, 2010). Moreover, there were differentials traits related to morphology, geography and host range that clearly justified the erection of a new species.

3) Another fact that supports the "resurrection" of *Hepatozoon felis* (Patton 1908) is based on coevolution of parasite and host (Criado-Fornelio *et al.*, 2003c). In this chapter we present for the first time a dating for the evolution of *Hepatozoon* sp in the mammalian carnivore lineage. The complete 18S rRNA gene sequences of *Theileria annae* (outgroup), *Hepatozoon* sp (bank vole 1), *H. felis* and several *H. canis* sequences (isolates Spain 1, Spain 2 and Curupira 1, 2 and 3) were aligned with ClustalW, and then analyzed with the MEGA 5 software to obtain a evolutionary history tree, which was based on the molecular clock hypothesis (Figure 6). The evolutionary history was inferred by using the Maximum Likelihood method based on the Kimura 2-parameter model. The landmark of separation of *Hepatozoon* from rodents and carnivores is dated at the Cretaceous–Tertiary (K-T) boundary (62 million years ago = MYA), as shown by the fossil record (Springer *et al.*, 2003). Such time landmark was introduced in the MEGA5 software for dating other nodes in the tree.

In summary, the results of the analysis indicate that:

- *H. felis* and the canid hepatozoon lineage diverged approximately 48 MYA (node 1).
- *H. americanum* and *H. canis* separated about 39 MYA (node 2).
- *H. canis* from dogs and European foxes separated 12 MYA (node 3)
- *H. canis* from European and South American foxes separated 6 MYA (node 4).

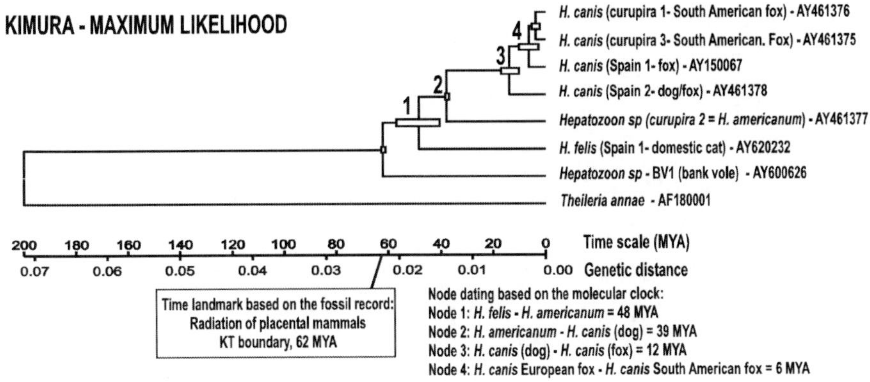

Figure 6. Phylogenetic relationships of *Hepatozoon* spp and *Babesia* (*Theileria*) *annae* based on DNA sequence variation of the 18S rRNA gene (MEGA 5 program). The Genbank® accession numbers are indicated after each name. The scale below indicates both time and genetic distance. The evolutionary history was inferred by using the Maximum Likelihood method (based on the Kimura two-parameter model). See the text of the chapter for more details on the analysis.

These data are more or less in agreement with the evolution of vertebrate hosts based on the fossil record or on studies employing on DNA sequence analysis. Thus, the last common ancestor of living carnivores may have existed as recently as 42 MYA (Polly *et al.*, 2006). However, other reports indicate that a common ancestor was already present about 55 MYA (Springer *et al.*, 2003). Therefore, the average of both hypotheses is 55 + 42 / 2 = 48.5, which is in quite good agreement with the timing obtained for node number 1.

The dating for node number 2 is 39 MYA, which could be correlated more or less with the first canines dated in the American fossil record (34 MYA), according to Tedford *et al.*, (2009).

The dating for node number 3 is 12 MYA, corresponding to the dichotomy of *Hepatozoon* from dogs (genotype Spain 2) and foxes (genotype Spain 1). This is in very good agreement with the timing of separation of foxes and dogs according to Tedford *et al.*, (2009). Indeed, the node number 4 may be correlated with the separation of European and South American foxes (genotypes Curupira 1 and 2), which dates back to 6 MYA according to the molecular clock hypothesis, and 5 MYA based on the fossil record (Tedford *et al.*, 2009).

Additional evidence for the relative accuracy of molecular dating is based on the fact that according to this method, the separation of *Theileria annae* and *Hepatozoon* occurred in the Jurassic period, 200 MYA (see Figure 6). This is in agreement with the idea that ticks appeared during the Mesozoic Era, as pointed out by Balashov (1989). Thus, it seems that in the primitive Apicomplexan arthropod-borne group, the divergence between two possible modes of transmission (one linked to ingestion, another to blood feeding) occurred in the early Mesozoic.

In conclusion, it is evident that the separation of canine and feline hepatozoons may be traced back to the period when both groups diverged (48 MYA), according to the phylogenetic analysis performed in this study using 18S rRNA gene sequences. This fact helps to reinforce the validity of the species *H. felis*.

Finally we must emphasize that the reapplication of the species denomination *Hepatozoon felis* is a conservative approach, since there is no need to erect new species. The evidence currently available suggests that name proposed by Patton (1908) for the hepatozoons of domestic cats has proven to be valid and thus, should be maintained (for isolates closely related to GenBank® accession numbers AY620232 and AY628681, as stated above), in agreement with the International code of Zoological Nomeclature (ICZN).

7. Control and Preventive Measures against Feline Hepatozoonosis

There are no vaccines available for feline hepatozoonosis (Bowman *et al.*, 2001; Baneth, 2011). Alternative preventive measures may be adopted against the arthropod vector. Concerning the vector of *H. canis* (and possibly *H. felis*), tick infestations are common and elimination can be expensive and time consuming. Effective tick control is difficult to achieve, because of the abundance of potential alternative hosts in the tick life cycle. Eradication is not feasible except on isolated locations such as islands. The use of acaricides in the environment (where tick infection usually occurs) is not exempt of risks (toxicity, residues in animal products and pollution). Treatment on animals by pour-on or spot-on procedures are therefore preferred (Jongehan and Uilenberg, 1994). Many effective host-targeted tick control agents exist, several of which also possess activity against other parasites (Blagburn and Dryden, 2009). Other strategies against ticks of cattle have been pointed out (e.g. anti-tick vaccines or looking for resistant hosts), but these are very difficult to apply in wild felines. Tick vaccines are complicated in the case of *Riphicephalus* species, since these are frequently three-host ticks and immunization of resevoirs may not be feasible.

The problem for other feline hepatozoons such as *H. felis* is the lack of knowledge on the vector. Both ticks and fleas have been pointed out as the putative invertebrate hosts (Criado-Fornelio, *et al.*, 2006 and Bortoli *et al.*, 2011). Recent data suggest that ticks are much more likely candidates, as shown in the present chapter. However, it must be considered that predation is likely to be the cause of some *Hepatoozon* infections (Smith, 1996; Salakij *et al.*, 2008 and 2010). If the ingestion of prey is a common way of infection, it would be very difficult to undertake any preventive action to protect wild animals (or domestic cats from rural areas with freedom for roaming and eating prey in forested regions) against hepatozoonosis.

Treatment of feline hepatozoonosis has seldom been reported. Since hepatoozoonosis in cats is usually subclinical, there is no standard chemotherapeutic protocol available. Van Amstel (1979) reported that a treatment with oxytetracycline and primaquine cured some infected animals in South Africa. Baneth *et al.*, (1998a) showed that treatment with doxycycline (5 mg/kg for 10 days) improved the condition of diseased cats.

CONCLUSION

Tick-borne hematozoans such as *Hepatozoon* spp are now considered as emergent pathogens. Hepatozoonosis is not a threat for the domestic cat, but some endangered feline species may be at risk, due to additional stress factors (such as climatic changes associated to global warming). This is more evident if concomitant pathogens (such as *Mycoplasma* spp or diverse viral diseases) are present. Since transmission might be linked to predation, preventive measures are very difficult to undertake, especially for wild felids. Moreover, the lack of knowledge on the distribution and ecology of *H. felis* (as well as its arthropod vectors) impairs the development and implementation of preventive and control measures at local, regional and national levels.

These facts emphasize the urgent need of additional research efforts on the epizootiology (studies on the geographic distribution of *H. felis* and the elucidation of the arthropod vectors associated to its transmission) as well as on the chemotherapy of feline hepatozoonosis.

Acknowledgments

I am deeply indebted to Dr. Jeronimo Ruas for his critical reading of the manuscript. I also wish to thank the veterinarians of the SMP animal shelter for their invaluable help in the collection of animal samples.

References

Abd-Rani, P.A., Irwin, P.J., Coleman, G.T., Gatne, M., Traub, R.J., (2011). A survey of canine tick-borne diseases in India. *Parasites and Vectors,* 4:141.

Allen, K.E., Li, Y., Kaltenboeck, B., Johnson, E.M., Reichard, M.V., Panciera, R.J., Little, S.E., (2008). Diversity of *Hepatozoon* species in naturally infected dogs in the southern United States. *Veterinary Parasitology,* 154, 220-225.

Allen, K.E., Yabsley, M.J., Johnson, E.M., Reichard, M.V., Panciera, R.J., Ewing, S.A., Little, S.E., (2011). Novel *Hepatozoon* in vertebrates from the southern United States. *The Journal of Parasitology,* 97, 648-653.

André, M.R., Adania, C.H., Teixeira, R.H., Vargas, G.H., Falcade, M., Sousa, L., Salles, A.R., Allegretti, S.M., Felippe, P.A., Machado, R.Z., (2010). Molecular detection of *Hepatozoon* spp. in Brazilian and exotic wild carnivores. *Veterinary Parasitology, 173*, 134-138.

Averbeck, G.A., Bjork, K.E., Packer, C., Herbst, L., (1990). Prevalence of hematozoans in lions (*Panthera leo*) and cheetah (*Acinonyx jubatus*) in Serengeti National Park and Ngorongoro Crater, Tanzania. *The Journal of Wildlife Diseases,* 26, 3942-3944.

Balashov, I.S., 1989. The coevolution of ixodid ticks and terrestrial vertebrates. *Parazitologiia,* 23, 457-468.

Baneth, G., Lavy, E., Presentey, B.Z., (1995). *Hepatozoon* sp parasitemia in a domestic cat. *Feline Practitioner,* 23, 10-12.

Baneth, G., Aroch, I., Tal, N., Harrus, S., (1998a). *Hepatozoon* species infection in domestic cats: a retrospective study. *Veterinary Parasitology,* 79,123-133.

Baneth, G., Shkap, V., Samish, M., Pipano, E., Savitsky, I., (1998b). Antibody response to *Hepatozoon canis* in experimentally infected dogs. Veterinary *Parasitology,* 74, 299-305.

Baneth, G., (2003). Disease risk for the travelling pet: hepatozoonosis. *In Practice,* 25, 272-277.

Baneth, G., Shkap, V., (2003). Monozoic cysts of *Hepatozoon canis.* The *Journal of Parasitology*, 89, 379-381.

Baneth, G., Kewnny, M.J., Tasker, S., Anug, Y., Shkap, V., Levy, A., Shaw, S.E., (2004). Infection with a proposed new subspecis of *Babesia canis*, *Babesia canis* subsp. *presentii,* in domestic cats. *Journal of Clinical Microbiology*, 42, 99-105.

Baneth, G., Samish, M., Shkap, V., (2007). Life cycle of *Hepatozoon canis* (Apicomplexa: Adeleorina: Hepatozoidae) in the tick *Riphicephalus sanguineus* and domestic dog (*Canis familiaris*). *The Journal of Parasitology,* 93, 283-299.

Baneth, G., (2011). Perspectives on canine and feline hepatozoonosis. *Veterinary Parasitology,* 181, 3-11.

Beaufils, J.P., (2004). Un cas d'hepatozoonose chez le chat. *Le nouveau praticien veterinaire,* 547, 39.

Berrada, Z.L., Telford, S.R., (2009). Burden of tick-borne infections on American companion animals. *Topics on Companion Animal Medicine,* 24, 175-181.

Bhadury, P., Austen, M., Bilton, D., Lambshead, P.J., Rogers, A., Smerdon, G., (2006). Developmemnt and evaluation of a DNA-barcoding approach for the rapid identification of nematodes. *Marine Ecology Progress Series,* 320, 1-9.

Blagburn, B.L, Dryden, M.W., (2009). Biology, treatment, and control of flea and tick infestations. *Veterinary Clinics North American Small Animals Practitioner*, 39,1173-1200.

Bortoli, C.P., André, M.R., Braga, M.S, Machado, R.Z., (2011). Molecular characterization of *Hepatozoon* sp. in cats from São Luís Island, Maranhão, Northeastern Brazil. *Parasitology Research, 109*, 1189-1192.

Bosman, A.M., Oosthuizen, M.C., Peirce, M.A., Venter, E.H., Penzhorn, B.L., (2010). *Babesia lengau* sp. nov., a novel *Babesia* species in cheetah (*Acinonyx jubatus*, Schreber, 1775) populations in South Africa. *Journal of Clinical Microbiology,* 48, 2703-2708.

Bowman, D.D., Hendrix, D.M., Lindsay, D.S., Barr, S.C., (2001). Hemogregarines. In: *Feline Clinical Parasitology*. Ed, Iowa State University Press, Ames, Iowa, pp 41-42.

Carneiro, M.P.M., (2007). Ocorrência de infecções por *Babesia* spp e *Hepatozoon* spp em gatos domésticos (*Felis domesticus*) do estado de São Paulo e do distrito federal. Master Dissertation, College of Veterinary Medicine and Zootechny, *Universidade Estadual Paulista* (Sao Paulo, Brazil), pp. 1-71.

Criado-Fornelio, A., Martinez-Marcos, A., Buling-Saraña, A., Barba-Carretero, J.C., (2003a). Presence of *Mycoplasma haemofelis*, *Mycoplasma haemominutum* and piroplasmids in cats from southern Europe: a molecular study. *Veterinary Microbiology,* 93, 307-317.

Criado-Fornelio, A., Martinez-Marcos, A., Buling-Saraña, A., Barba-Carretero, J.C., (2003b). Molecular studies on *Babesia*, *Theileria* and *Hepatozoon* in southern Europe. Part I. Epizootiological aspects. *Veterinary Parasitology,* 113, 189-201.

Criado-Fornelio, A., Martinez-Marcos, A., Buling-Saraña, A., Barba-Carretero, J.C., (2003c). Molecular studies on *Babesia*, *Theileria* and *Hepatozoon* in southern Europe. Part II. Phylogenetic analysis and evolutionary history. *Veterinary Parasitology,* 114, 173-194.

Criado-Fornelio, A., Gónzalez-del-Río, M.A., Buling-Saraña, A., Barba-Carretero, J.C., (2004). The "expanding universe" of piroplasms. *Veterinary Parasitology,* 119, 337-345.

Criado-Fornelio, A., Ruas, J.L., Casado, N., Farias, N.A., Soares, M.P., Müller, G., Brumt, J.G., Berne, M.E., Buling-Saraña, A., Barba-Carretero, J.C., (2006). New molecular data on mammalian *Hepatozoon* species (Apicomplexa: Adeleorina) from Brazil and Spain. *The Journal of Parasitology,* 92, 93-99.

Criado-Fornelio, A., Rey-Valeiron, C., Buling, A., Barba-Carretero, J.C., Jefferies, R., Irwin, P., (2007a). New advances in molecular epizootiology of canine hematic protozoa from Venezuela, Thailand and Spain. *Veterinary Parasitology,* 144, 261-269.

Criado-Fornelio, A., Buling, A., Cunha-Filho, N.A., Ruas, J.L., Farias, N.A., Rey-Valeiron, C., Pingret, J.L., Etievant, M., Barba-Carretero, J.C., (2007b). Development and evaluation of a quantitative PCR assay for detection of *Hepatozoon* sp. *Veterinary Parasitology,* 150, 352-356.

Criado-Fornelio, A., Buling, A., Casado, N., Gimenez, C., Ruas, J.L., Wendt, L., Farias, N.A., Pinheiro, M., Rey-Valeiron, C., Barba-Carretero, J.C., (2009a). Molecular characterization of arthropod-borne hematozoans in wild mammals from Brazil, Venezuela and Spain. *Acta Parasitologica,* 54, 187-193.

Criado-Fornelio, A., Buling, A., Pingret, J.L., Etievant, M., Boucraut-Baralon, C., Alongi, A., Agnone, A., Torina, A., (2009b). Hemoprotozoa of domestic animals in France: prevalence and molecular characterization. *Veterinary Parasitology.* 159, 73-76.

East, M.L., Wibbelt, G., Lieckfeldt, D., Ludwig, A., Goller, K., Wilhelm, K., Schares, G., Thierer, D., Hofer, H., (2008). A *Hepatozoon* species genetically distinct from *H. canis* infecting spotted hyenas in the Serengeti ecosystem, Tanzania. *Journal of Wildlife Diseases,* 44, 45-52.

Eizirik, E., Murphy, W.J., Koepfli, K.P., Johnson, W.E., Dragoo, J.W., Wayne, R.K., O'Brien, S.J., (2010). Pattern and timing of diversification of the mammalian order Carnivora inferred from multiple nuclear gene sequences. *Molecular Phylogenetics and Evolution,* 56, 49-63.

Ewing, G. O., (1977). Granulomatous cholangiohepatitis in a cat due to a protozoan parasite resembling *Hepatozoon canis. Feline Practice,* 7, 37-40.

Fishman, Z., Gonen, L., Harrus, S., Strauss-Ayali, D., King, R., Baneth, G., (2004). A serosurvey of *Hepatozoon canis* and *Ehrlichia canis* antibodies in wild red foxes (*Vulpes vulpes*) from Israel. *Veterinary Parasitology,* 119, 21-26.

Giannitti, F., Diab, S.S., Uzal, F.A., Fresneda, K., Rossi, D., Talmi-Frank, D., Baneth, G., (2012). Infection with a *Hepatozoon* sp. closely related to *Hepatozoon felis* in a wild Pampas gray fox (*Lycalopex -Pseudalopex - gymnocercus*) co-infected with canine distemper virus. *Veterinary Parasitology,* in press.

Gonen, L., Strauss-Ayali, D., Shkap, V., Vincent-Johnson, N., Macintire, D.K., Baneth G., (2004). An enzyme-linked immunosorbent assay for antibodies to *Hepatozoon canis. Veterinary Parasitology.* 122, 131-139.

Götsch, S., Leschnik, M., Duscher, G., Burgstaller, J.P., Wille-Piazzai, W., Joachim, A., (2009). Ticks and haemoparasites of dogs from Praia, *Cape Verde. Veterinary Parasitology,* 166, 171-174.

Harrus, S., Baneth, G., (2005). Drivers for the emergence and re-emergence of vector-borne protozoal and bacterial diseases. International *Journal for Parasitology,* 35,1309-1318.

Hopla, C.A, Durden, L.A., Keirans, J.E., (1994). Ectoparasites and classification. *Office Internationale des épizooties,* 13, 985-1017

Inokuma, H., Okuda, M., Ohno, K., Shimoda, K., Onishi, T., (2002). Analysis of the 18S rRNA gene sequence of a *Hepatozoon* detected in two Japanese dogs. *Veterinary Parasitology,* 106, 265-71.

Jittapalapong, S., Rungphisutthipongse, O., Maruyama, S., Schaefer, J.J., Stich, R.W., (2006). Detection of *Hepatozoon canis* in stray dogs and cats in Bangkok, Thailand. *Annals of the New York Academy of Sciences,* 1081, 479-488.

Jongehans, F., Uilenberg, G. (1994). Ticks and control methods. *Revue Scientifique et Technique – Office International des Epizooties,* 13, 1201-1226.

Joyner, P.H., Reichard, M.V., Meinkoth, J.H., Milne, V.E., Confer, A.W., Kocan, A.A, Hoover, J.P., (2007). Experimental infection of domestic cats (*Felis domesticus*) with *Cytauxzoon manul* from Pallas' cats (*Otocolobus manul*). *Veterinary Parasitology,* 146, 302-306.

Klopfer, U., Nobel, T.A., Neumann, F., (1973). *Hepatozoon*-like parasite (schizonts) in the myocardium of the domestic cat. *Veterinary Pathology,10*, 185-190.

Kubo, M, Miyoshi N, Yasuda N., (2006). Hepatozoonosis in two species of Japanese wild cat. *The Journal of Veterinary Medical Science,* 68, 833-887.

Kubo, M, Jeong, A., Kim, S.I., Kim, Y.J., Lee, H., Kimura, J., Agatsuma, T., Sakai, H., Yanai, T. (2010). The first report of *Hepatozoon* species infection in leopard cats (*Prionailurus bengalensis*) in Korea. *The Journal of Parasitology,* 96, 437-439.

Li, Y., Wang, C., Allen, K.E., Little, S.E., Ahluwalia, S.K., Gao, D., Macintire, D.K., Blagburn, B.L., Kaltenboeck, B., (2008). Diagnosis of canine *Hepatozoon* spp. infection by quantitative PCR. *Veterinary Parasitology,* 157, 50-58.

McCully, R.M., Basson, P.A., Bigalke, R.D., De Vos, V., Young, E., (1975). Observations on naturally acquired hepatozoonosis of wild carnivores and dogs in the Republic of South Africa. *Onderstepoort Journal of Veterinary Research*, 42, 117-133.

Mathew, J.S., Saliki, J.T., Ewing, S.A., Lehenbauer, T.W., Panciera, R.J., Malayer, J.R., Cummings, C.A., Kocan, A.A., (2001). An indirect enzyme-linked immunosorbent assay for diagnosis of American canine hepatozoonosis. *Journal of Veterinary Diagnostic Investigation,* 13, 17-21.

Mendes, F., Crissiuma, A.L., Gershony, L.C., Willi, L.M., Paiva, J.P., Guerrero, J., Labarthe, N., (2011). Characterization of ectoparasites in an urban cat (*Felis catus* Linnaeus, 1758) population of Rio de Janeiro, Brazil. *Parasitology Research,* 108, 1431-1435.

Mercer, S.H., Jones, L.P., Rappole, J.H., Twedt, D., Laack, L.L., Craig, T.M., (1988). *Hepatoozon* sp in wild carnivores in Texas. *Journal of Wildlife Diseases,* 24, 574-576.

Metzger, B., dos Santos-Paduan, K., Rubini, A.S., de Oliveira, T.G., Pereira, C., O'Dwyer, L.H., (2008). The first report of *Hepatozoon* sp. (Apicomplexa: Hepatozoidae) in neotropical felids from Brazil. *Veterinary Parasitology,* 152, 28-33.

Murata, T., Inoue, M., Tateyama, S., Taura, Y., Nakama, S. (1993). Vertical transmission of *Hepatozoon canis* in dogs. *The Journal of Veterinary Medical Science,* 55, 867-868.

Ortuño, A., Castellà, J., Criado-Fornelio, A., Buling, A., Barba-Carretero, J.C., (2008). Molecular detection of a *Hepatozoon* species in stray cats from a feline colony in North-eastern Spain. *Veterinary Journal,* 177, 134-5.

Oyamada, M., Davoust, B., Boni, M., Dereure, J., Bucheton, B., Hammad, A., Itamoto, K., Okuda, M., Inokuma, H., (2005). Detection of *Babesia canis rossi, B. canis vogeli,* and *Hepatozoon canis* in dogs in a village of eastern Sudan by using a screening PCR and sequencing methodologies. *Clinical Diagnostic Laboratory Immunology,* 12, 1343-1346.

Patton, W.S., (1908). The haemogregarines of mammals and reptiles. *Parasitology,* 1, 318-321.

Pawar, R.M., Poornachandar, A., Srinivas, P., Rao, K.R., Lakshmikantan, U., Shivaji, S., (2012). Molecular characterization of *Hepatozoon* spp. infection in endangered Indian wild felids and canids. *Veterinary Parasitology,* in press.

Perez, R.R., Rubini, A.S., O'Dwyer, L.H., (2004). The firtst report of *Hepatozoon* sp (Apicomplexa: Hepatozoidae) in domestic cats from Sao Paulo state, Brazil. *Parasitology Research,* 97, 91-95.

Polly, P.D., Wesley-Hunt, G.D., Heinrich R.E, Davis, G., Houde, P., (2006). Earliest known carnivoran auditory bulla and support for a recent origin of crowngroup Carnivora (Eutheria, Mammalia). *Palaeontology,* 49, 1019–1027.

Reichard, M.V., Van Den Bussche, R.A., Meinkoth, J.H., Hoover, J.P., Kocan, A.A., (2005). A new species of *Cytauxzoon* from Pallas' cats caught in Mongolia and comments on the systematics and taxonomy of piroplasmids. *The Journal of Parasitology,* 91, 420-426.

Rubini, A.S., Dos Santos-Paduan, K., Perez, R.R., Ribolla, P.E., O'Dwyer, L.H., (2006). Molecular characterization of feline *Hepatozoon* species from Brazil. *Veterinary Parasitology,* 137, 168-171.

Salakij, C., Salakij, J., Narkkong, N., Sirinarumitr, T., Pattanarangsan, R., (2008). Hematologic, cytochemical, ultrastructural, and molecular findings of *Hepatozoon*-infected flat-headed cats (*Prionailurus planiceps*). *Veterinary Clinical Pathology,* 37, 31-41.

Salakij, C., Sirinarumitr, T., Tongthainun, D., (2010). Molecular characterization of *Hepatozoon* species in a Leopard Cat (*Prionailurus bengalensis*) from Thailand. *Veterinary Clinical Pathology,* 39, 199-202.

Sakuma, M., Nishio, T., Nakanishi, N., Izawa, M., Asari, Y., Okamura, M., Shimokawa-Miyama, T., Setoguchi, A., Endo, Y., (2011). A case of Iriomote cat (*Prionailurus bengalensis iriomotensis*) with *Hepatozoon felis* parasitemia. *Journal of Veterinary Medical Science,* 73, 1381-1384.

Smith, T. G. 1996. The genus *Hepatozoon* (Apicomplexa: Adeleina). Journal of Parasitology, 82**,** 565-585.

Solano-Gallego, L., Baneth, G. (2011). Babesiosis in dogs and cats - expanding parasitological and clinical spectra. *Veterinary Parasitology,* 181, 48-60.

Spolidorio, M.,G., Labruna, M.B., Zago, A.M., Donatele, D.M., Caliari, K.M., Yoshinari, N.H., (2009). *Hepatozoon canis* infecting dogs in the State of Espírito Santo, southeastern Brazil. *Veterinary Parasitology,* 163, 357-361.

Springer, M.S., Murphy, W.J., Eizirik, E., O'Brien, S.J., (2003). Placental mammal diversification and the Cretaceous-Tertiary boundary. *Proceedings of the National Academy of Sciences of the U.S.A.,* 100, 1056-1061.

Tabar, M.D., Altet, L., Francino, O., Sánchez, A., Ferrer, L., Roura, X., (2008). Vector-borne infections in cats: molecular study in Barcelona area (Spain). *Veterinary Parasitology,* 151, 332-336.

Tabar, M.D., Francino, O., Altet, L., Sánchez, A., Ferrer, L., Roura, X., (2009). PCR survey of vectorborne pathogens in dogs living in and around Barcelona, an area endemic for leishmaniasis. *The Veterinary Record,* 24, 112-116.

Tedford, R.H., Wang, X., Taylor, B.E., (2009). Phylogenetic Systematics of the North American Fossil Caninae (Carnivora: Canidae). *Bulletin of the American Museum of Natural History,* 325, 1-218.

Telford, S.R., Wozniak, E.J, Butler, J.F., (2001). Haemogregarine specificity in two communities of Florida snakes, with descriptions of six new species of *Hepatozoon* (Apicomplexa: Hepatozoidae) and a possible species of *Haemogregarina* (Apicomplexa: Haemogregarinidae). *The Journal of Parasitology,* 87, 890-905.

Van Amstel, S., (1979). Feline hepatozoonosis. *Journal of the South African Veterinary Association,* 50, 215-216.

Vargas-Hernández, G., André, M.R., Munhoz, T.D., Faria, J.M., Machado, Z.M., Tinucci-Costa, M., (2011). Molecular characterization of *Hepatozoon canis* in dogs from Colombia. *Parasitology Research,* 110, 489-492.

Vojta, L., Mrljak, V., Curković, S., Zivicnjak, T., Marinculić, A., Beck, R., (2009). Molecular epizootiology of canine hepatozoonosis in Croatia. *International Journal for Parasitology,* 39, 1129-1136.

Wenyon, C.M., (1926). *Protozoology: a manual for medical man veterinarians and zoologists.* William Wood, New York, 1085-1095.

Widmer, C.E., (2009). Perfil sanitario des onças-pintadas (*Panthera onca*) da vida libre no Pantanal de Sul de Matto Grosso do Sul - Brasil. *PhD Dissertation,* College of Veterinary Medicine, University of Sao Paulo, pp. 1-78.

Revised by Dr. Jeronimo Ruas, Laboratorio Regional de Diagnostico, Faculdade de Veterinária (LRD). Universidade Federal de Pelotas, P.O. Box 354 Pelotas, Brazil.

In: Felines
Editors: R. García and R. Jiménez
ISBN 978-1-62257-005-8

Chapter 4

THE ROLE OF CATS IN *TOXOPLASMA GONDII* INFECTION IN SPAIN REVIEW

Oscar Cabezón,[1] Javier Millán,[1] Ignacio García-Bocanegra[2], and Sonia Almería[3,4]
[1]Servei d'Ecopatologia de Fauna Salvatge (SEFaS) (Wildlife Diseases Research Group), Departament de Medicina i Cirurgia Animals, Universitat Autònoma de Barcelona, Bellaterra, Spain
[2]Departamento de Sanidad Animal, Facultad de Veterinaria, Universidad de Córdoba (UCO), Campus Universitario de Rabanales, Córdoba, Spain
[3]Departament de Sanitat i d'Anatomia Animal, Facultat de Veterinària, Universitat Autònoma de Barcelona, Bellaterra, Barcelona, Spain
[4]Centre de Recerca en Sanitat Animal (CReSA), UAB-IRTA, Campus de la Universitat Autònoma de Barcelona, Bellaterra, Barcelona, Spain

ABSTRACT

Toxoplasma gondii is a zoonotic intracellular protozoan parasite of worldwide distribution. The parasite is a main cause of abortion in small ruminants and humans when primary infection occurs during pregnancy. Felids, with particular importance of cats, are the definitive hosts, excreting oocysts in the faeces. Humans and virtually all warm-blooded species can be intermediate hosts and can become infected by ingestion of food and water contaminated with sporulated *T. gondii* oocysts, by consumption of tissue cysts in infected animal tissues, or congenitally. In the last decade, several studies of seroprevalence of *T. gondii* in numerous domestic and wild animal species (cats, dogs, pigs, wild felids, red deer and other cervids, wild boars, wild rabbits, wild birds and dolphins) have confirmed that *T. gondii* infection is widespread in Spain. Cats have a prominent role in the transmission of the infection. Recent studies in both domestic and wild cats from different areas of Spain have shown elevated seroprevalence levels, e.g: 45% of 220 domestic cats from Barcelona (Catalonia, North East Spain), and 32.3% of 585 domestic cats from Madrid and La Rioja (Central-North Spain). The highest seroprevalence has been observed in feral cats from Mallorca (Balearic Islands) (84.7% of 59 cats), being one of the highest reported worldwide in this species and the highest observed in Europe to date. High seroprevalence has also been observed in wild felids, such as Iberian lynx (81.5% of 27 animals, and 62.8% of 129 animals in two separate studies), and in European wildcats (50% of 6 animals). Furthermore, the presence of cats has been reported as one of the main factors for *T. gondii* seroprevalence in other domestic species, such as pigs and small ruminants in Spain. Seroprevalence levels in pregnant women in Spain have been reported to range from 18.8% in 2,929 pregnant women in Salamanca to 28.6% in 16,362 pregnant women in Barcelona. Since seropositive felids are likely to have already shed *T. gondii* oocysts in the environment, the high rate of *T. gondii* seroprevalence in cats can have important implications for conservation and animal infection and public health in Spain.

Introduction

Toxoplasma gondii is an intracellular protozoan (Phylum Aplicomplexa) of worldwide distribution. *Toxoplasma gondii* was first identified in 1908 in a North African rodent (*Ctenodactylus gondi*) by Nicolle & Manceaux (1909). Domestic and free-ranging felids are the only definitive hosts for *T. gondii*. In infected cats, oocysts are produced and shed in the feces. In fact, cats eliminate oocysts of *T. gondii* in feces after its sexual replication usually for a brief period after primary infection but can excrete millions of oocysts per gram of feces. The oocysts are extremely resistant into the environment where can survive infective up to 2 years depending the temperature or humidity conditions (Dubey, 2009a).

The definitive and intermediate host may be infected by ingestion of water or food contaminated with oocysts, by ingestion of tissue cysts or by vertical transmission (Dubey and Beattie, 1988; Dubey 2009a). Virtually all warm-blooded hosts, including mammals, birds and humans, can be intermediate hosts of *T. gondii* in which the parasite remains as bradyzoites in slow-growing tissue cysts that persist chronically in tissues (mainly muscle and nervous system) (Dubey et al., 1997).

Toxoplasmosis can have severe clinical manifestations in both humans and animals. In humans, a correlation between severe toxoplasmosis and immune-suppression that can lead to the death of the host has been well documented. Another severe clinical manifestation of toxoplasmosis has been described when *T. gondii* infects a pregnant woman during pregnancy, when the parasite can cause abortion or congenital disease in the foetus (Jones et al., 2002; Remington et al., 2006).

Toxoplasma gondii is known to infect a wide range of domestic, livestock and wild species, causing economic losses worldwide. However, clinical toxoplasmosis depends on the species affected. *T. gondii* normally appears as a mild and transient disease, establishing long-term latent infection in the host. However, in cats toxoplasmosis is characterized by mild enteritis, with diarrhea, during 1 to 2 weeks after primary infection. Even so, in some cases felids can develop fatal clinical toxoplasmosis for example when cats suffer from immune-suppression (co-infection with Feline Immunodeficiency Virus, Feline Infectious Peritonitis or Feline Leukemia Virus among others) (Dubey, 2009a). Livestock species such as sheep are very susceptible to *T. gondii*, which causes abortion and neonatal mortality in sheep including in Spain (Pereira-Bueno et al., 2004; Moreno et al., 2012).. Other species such as marsupials or new-world primates also suffer severe disease that can be fatal. An epizootic of toxoplasmosis among captive Bennett's wallabies (*Macropus rufogriseus*) was recently reported in Spain (Bermúdez et al., 2009). The possible sources of infection were feral cats or reactivation of a latent infection and may constitute a risk of infection for humans since new dietary and cooking practices are being introduced to Spain (Bermudez et al., 2009). In contrast, some species such as equids or cattle are known to be toxoplasmosis resistant (Dubey and Beattie, 1998; Dubey and Jones, 2008).

In general, many lesional and clinical symptoms associated with *T. gondii* in wildlife species are not well known. Symptomatic toxoplasmosis has been reported in several free-ranging wild species such as the red-necked wallaby (*Macropus rufogriseus*) (Johnson et al., 1989; Bermúdez *et al.*, 2009), the Pallas's cat (*Otocolobus manul*) (Basso et al., 2005), the red fox (*Vulpes vulpes*) (Dubey et al., 1990), the cheetah (*Acinonyx jubatus*) (Lloyd y Stidworthy, 2007), and the white-tailed deer (*Odocoileus virginianus*) (Dubey et al., 2008).

In Spain, other than the clinical problems associated with sheep abortion and the case report in wallabies above, systemic toxoplasmosis has been described in a Pyrenean chamois (*Rupicapra pyrenaica*) (Marco et al., 2009), in striped dolphins (*Stenella coeruleoalba*) affected with Dolphin Morbillivirus (Domingo et al., 1992); in a pregnant Risso's dolphin (*Grampus griseus*) from the Spanish Mediterranean coast (Resendes et al., 2002), and in several captive exotic species such as slender-tailed meerkats (*Suricata suricatta*), golden lion tamarins (*Leontopithecus rosalia*) and ring-tailed lemur (*Lemur catta*) (Juan-Sallés et al., 1997; Juan-Sallés et al., 1998; Juan-Sallés et al., 2011).

Serological surveys have confirmed that *T. gondii* infection is widespread in numerous domestic and wild animal species in Spain, including cats, dogs, ferrets (*Mustela putorius furo*), pigs, sheep, cattle, equids (horse, donkey and mule), Iberian lynx (*Lynx pardinus*), European wildcat (*Felis silvestris*), red fox (*Vulpes vulpes*), wolf (*Canis lupus*), Eurasian badger (*Meles meles*), stone marten (*Martes foina*), pine marten (*Martes martes*), Eurasian otter (*Lutra lutra*), polecat (*Mustela putorius*), European genet (*Genetta genetta*), Egyptian mongooses (*Herpestes ichneumon*), red deer (*Cervus elaphus*), roe deer (*Capreolus capreolus*), fallow deer (*Dama dama*), Spanish ibex (*Capra pyrenaica*), chamois (*Rupicapra pyrenaica*), barbary sheep (*Ammotragus lervia*), mouflon (*Ovis ammon*), wild boar (*Sus scrofa*), wild rabbit (*Oryctolagus cuniculus*), various species of wild birds, and dolphins (Gauss et al., 2003; Miró et al., 2004; Almería et al., 2004; Cabezón et al., 2004; Gauss et al., 2005; Gauss et al., 2006; Sobrino et al., 2007; Gamarra et al., 2008; Millán et al., 2009a; García-Bocanegra et al., 2010a; 2010b; Cabezón et al., 2010b; Closa-Sebastià et al., 2011; Molina-López et al., 2012; Cabezón et al., 2012; García-Bocanegra et al., 2012a; 2012b).

2. Seroprevalence Studies of *Toxoplasma gondii* in Felids in Spain

Cats are important in the epidemiology of *T. gondii* infection because they are the only hosts that can excrete the environmentally resistant oocysts. Reports on seroprevalence are of interest because seropositive cats are likely to have already shed *T. gondii* oocysts in the environment (Dubey, 2009a). In addition, surveys of *T. gondii* infection in free-ranging felids can provide estimates of environmental contamination and circulation of *T. gondii* in domestic and wild ecosystems.

2.1. Domestic Cat (*Felis silvestris catus*)

Until recently only single study (Aparicio-Garrido et al., 1972) had reported on the prevalence of *T. gondii* antibodies in sera of 55 of 101 cats (54.4%) from Madrid, central Spain. Fortunately, further studies of seroprevalence of *T. gondii* in cats have been reported in the last decade in Spain (Gauss et al., 2003, Miró et al., 2004; Millán et al., 2009a, Millán et al., 2009b) (Table 1).

In the first of these recent studies, antibodies to *T. gondii* were determined in serum samples from 220 cats from Barcelona, North-East Spain, characterized by a Mediterranean climate. The technique used was the modified agglutination test (MAT). Antibodies to *T. gondii* were found in 99 (45%) of 220 cats. The seropositivity (MAT 1:25 or more) was significantly higher in adult (> or = 1 yr old, 49.7% of 153) than in juvenile (< 1 yr old, 34.3% of 67) cats, in free-roaming (51.9% of 31) than in domiciled (34.8% of 89) cats, and in cats living in a group (community) of more than 5 cats (50.7% of 142) than in cats living alone (28.0% of 50) (Gauss et al., 2003).

The seroprevalence of *T. gondii* in cats varies, according to several factors, the type of cat (free-roaming vs. domestic), age of cats, method of serologic testing, and/or geographic location. In the study by Gauss et al. (2003), increased infection levels were observed in older cats in Barcelona, which were probably due to the fact that adult cats have more exposure to *T. gondii* in the course of their life time. In the same study, feral cats had the highest prevalence of infection. This observation agreed with many previous studies (Dubey 1973; Dubey et al., 2002). Feral cats living free in the streets could be more susceptible to exposure to oocysts from the environment and through ingestion of tissue cysts from intermediate hosts like wild birds or rodents. On the other hand, Gauss et al., (2003) observed moderate prevalence of *T. gondii* in cats living in houses or in apartments but with free access outdoors which suggested that those cats also had the opportunity to hunt for small prey and/or be exposed to oocysts. Even cats that lived in apartments and were kept strictly indoors had a considerable rate of infection (31.9%) in that study. The authors explained that contact with the parasite in those animals was probably due to the cats being fed raw or undercooked meat during gestation (Gauss et al., 2003).

In a second recent study on cats from central-North areas of Spain, in Madrid- La Rioja (Miró et al., 2004), the seroprevalence of *T. gondii* was analyzed in 585 cats using the indirect immunofluorescence antibody test (IFAT). This study detected antibodies in 189 (32.3%) individuals. As in the study of Barcelona, a higher prevalence was observed in adult cats (>6 months old, 36.8% of 443) compared to juvenile cats (<6 months old, 13.9% of 101). Miró et al., (2004) showed that seroprevalence was higher in male (45.3% of 128) than in female (32% of 153) stray cats, while in the study in Barcelona no statistically significant differences were observed between sexes.

Table 1. Seroprevalence levels and/or presence of *Toxoplasma gondii* in carnivorous species from Spain. CC: Cantabric Coastal región; CL: Castile and León; SC: Central Spain; CV: Comunitat Valenciana; AND: Andalusia; CAT: Catalonia; MAD: Madrid; AR: Aragón; LR: La Rioja; M: Mallorca

Species	Location	N	Seropositive (%)	Test	Reference
Domestic cat (*Felis silvestris catus*)	MAD	101	54.4	IFAT	Aparicio Garrido et al., 1972
	CAT	89	34.8	MAT	Gauss et al., 2003
	MAD	220	25.5	IFAT	Miró et al., 2004
	LR, MAD	18	44.4	PCR	Montoya et al., 2008
Feral cat (*Felis silvestris catus*)	CAT	31	51.9	MAT	Gauss et al., 2003
	MAD	317	36.9	IFAT	Miró et al., 2004
	LR, MAD	73	30.1	PCR	Montoya et al., 2008
	AND	25	52.0	MAT	Millán et al., 2009a
	M	59	84.7	MAT	Millán et al., 2009b
European wild cat (*Felis silvestris silvestris*)	CC, SC	6	50.0	MAT	Sobrino et al., 2007
Iberian lynx (*Lynx pardinus*)	AND	27	81.5	MAT	Sobrino et al., 2007
	AND	26	80.7	MAT	Millán et al., 2009b
	AND	48	44.0	LAT, IHA	Roelke et al., 2008
	AND	129	62.8	MAT	García-Bocanegra et al., 2010b

Species	Location	N	Seropositive (%)	Test	Reference
Dog (*Canis familiaris*)	M	46	58.7	MAT	Cabezón et al., 2010
	AND	28	53.0	MAT	Millán et al., 2009a
Wolf (*Canis lupus*)	CC, CL	32	46.5	MAT	Sobrino et al., 2007
Red fox (*Vulpes vulpes*)	CC, CL, SC, AND, AR	102	64.7	MAT	Sobrino et al., 2007
Eurasian badger (*Meles meles*)	CC, SC, AND	37	70.3	MAT	Sobrino et al., 2007
Stone marten (*Martes foina*)	CC, SC, CV	20	85.0	MAT	Sobrino et al., 2007
Pine marten (*Martes martes*)	CC	4	100.0	MAT	Sobrino et al., 2007
Eurasian otter (*Lutra lutra*)	CC, SC	6	100.0	MAT	Sobrino et al., 2007
Polecat (*Mustela putorius*)	CC, SC	4	100.0	MAT	Sobrino et al., 2007
Ferret (*Mustela putorius furo*)	CC	1	100.0	MAT	Sobrino et al., 2007
European genet (*Genetta genetta*)	CC, SC, AND, CV	21	61.9	MAT	Sobrino et al., 2007
Species	Location	N	Seropositive (%)	Test	Reference
Egyptian mongoose (*Herpestes ichneumon*)	SC, AND	22	59.1	MAT	Sobrino et al., 2007
Wild birds - Falconiformes	CAT, AND, SC, CL	602	23.8	MAT	Cabezón et al., 2011
Wild birds - Strigiformes	CAT, AND, SC, CL	259	44.0	MAT	Cabezón et al., 2011
Wild birds - Common raven (*Corvus corax*)	CAT	113	80.5	MAT	Molina-López et al., 2012
Wild birds – Carnivorous / Scavenger spp.	CAT, CL	200	6.0	PCR	Darwich et al., 2012

Different *T. gondii* strains have been associated with a wide range of clinical manifestations in infected species of animals and humans. To our knowledge, to date only, Montoya et al., (2008) have performed characterization of *T. gondii* genotypes in cats from Spain. In this study, fresh samples of brain from 103 Spanish cats with antibodies to *T. gondii* were analyzed for *T. gondii* DNA using nested-PCR. *Toxoplasma gondii* was detected in 47 (45.5%) of the cats analyzed. Further characterization of DNA from 46 cats using RFLP-PCR at the 3' and 5' ends of the SAG2 locus revealed that 12 (26%) isolates were Type I and 34 (74%) were Type II; no Type III were found.

2.2. Importance of the Lifestyle of Cats

Gauss et al. (2003) and Miró et al. (2004) reported a higher seroprevalence to *T. gondii* in free-roaming cats when compared with household cats in north-east and north-central Spain, respectively. All free-roaming cats studied in Barcelona by Gauss et al. (2003) lived in a community, and the study showed that community cats had a statistically significant higher prevalence of *T. gondii* infection than cats living alone. An one explanation was that cats living in a community likely share the same surroundings and thus have higher exposure to *T. gondii* in the environment. European and mixed breed cats, the most common breeds found as stray cats in the cities of Spain, had a higher prevalence of *T. gondii* infection than pure non-European breeds (Gauss et al., 2003). Purebred non-European cats are usually fed commercial canned food instead of raw or uncooked meat or live-prey and are generally kept indoors. These data suggest that living in a community, type of diet and free access to the outdoor environment are equally important as predisposing factors to the risk of *T. gondii* infection in cats in the Barcelona area (Gauss et al., 2003).

Most of the serological surveys in cats worldwide are from domestic cats in urban or semi-urban areas like the ones described above in Spain. There is little information from strictly wild living cats because these cats are more difficult to study. A feral cat is defined as a cat that is not attached to a particular household (Liberg et al., 2000). Such cats have no specific owner, are not subject to vaccination programs and do not receive prophylactic or curative treatments against parasites.

In a study in Andalusia (South of Spain), Millán et al. (2009a) found MAT (≥1:25) antibodies in 52% of 25 feral cats. Also, oocysts of *T. gondii* were detected in feces from 17% of these feral cats by direct immunofluorescence. In the latter study, both seroprevalence and active elimination of the parasite showed that feral cats have an active repetition role in the dissemination of the parasite in Southern Spain, increasing the probability of infection of other potential intermediary and/or definitive host species.

Similar data was recorded in another survey of antibodies to *T. gondii* in feral cat populations from Mallorca Island (Millán et al., 2009b). Mallorca is the largest of the Balearic Islands (Spain) located in the Mediterranean Sea. In the rural areas of the island, there is an abundant population of free-living feral cats. The study analyzed sera from 59 feral cats captured in baited traps during authorized predator control campaigns. Antibodies to *T. gondii* assayed by MAT were found in 50 (84.7%) of 59 cats. Seroprevalence (MAT, ≥1:25) was, as in the previous studies in Spain, significantly higher in adults (94.6%) than in juveniles (<6 months old; 40.0%). Interestingly, the seroprevalence of *T. gondii* observed in cats from Mallorca is one of the highest reported in this species worldwide and the highest reported in feral cats in Europe to date (reviewed by Jones and Dubey, 2010). It has been suggested that cats are more susceptible to *T. gondii* infection from eating tissues of infected animals rather that infection by oocysts (Dubey and Beattie, 1988). In this regard, Millán (2010) studied the diet of these cats and determined that rodents (*Rattus* sp., *Mus* sp., *Apodemus sylvaticus*) constituted the main prey item in feral cat´s year round diet in the countryside of Mallorca. To date, no studies on seroprevalence in rodents have been reported in Spain. It is possible that the Mediterranean climate, could favour survival of *T. gondii* oocysts shed by these feral cats increasing the possibility of infection of intermediary hosts such as rodents or livestock representing a public health threat. Riera et al. (1995) reported that 18% of AIDS patients in Mallorca and Ibiza developed cerebral toxoplasmosis.

2.3. Wild Felids

There are two wild felid species in Spain: the Iberian lynx (*Lynx pardinus*) and the European wildcat (*Felis silvestris silvestris*). The Iberian lynx is the most endangered felid species in the world (Nowell and Jackson, 1996) and the most endangered carnivore in Europe (IUCN, 2011) with approximately 250 individuals currently inhabiting only two isolated metapopulations in southern Spain (Sierra Morena and Doñana) (Palomares et al., 2002). The European wildcat (*Felis silvestris silvestris*) is categorized as vulnerable in Spain (Palomo and Gisbert, 2002), on the basis of a suspected population decline of > 30% over the last three generations. The wildcats are found in a wide variety of habitats, but are primarily associated with forests with a low density of humans. Rodents and rabbits are their staple diet and they also eat birds. In addition, wildcats have also been observed to show scavenger behaviour (Nowell and Jackson 1996, Sunquist and Sunquist 2002). The study of *T. gondii* infection in wild felids in Spain is of interest because they can act as definitive hosts and because the infection by *T. gondii* can affect endangered species that suffer from a high degree of inbreeding, such as the Iberian lynx.

2.3.1. Iberian Lynx (Lynx pardinus)

There have been several studies of *T. gondii* seroprevalence in Iberian lynx (Rodríguez and Carbonell, 1998; Sobrino et al., 2007; Roelke et al., 2008; Millán et al., 2009b; García-Bocanegra et al., 2010b). In the most recent studies, Sobrino et al., (2007) and Millán et al., (2009b) observed a higher seroprevalence (>80%) than Roelke et al., (2008) that reported seroprevalence of 40% in 48 animals (Table 1). The most extensive study in Iberian lynx, however, has been performed by García-Bocanegra et al., (2010b). In this study, seroprevalence and associated risk factors for *T. gondii* infection in a large population of Iberian lynx (129 animals from 2005 to 2009) were determined and antibodies to *T. gondii* were found in 62.8% Iberian lynx. Seroprevalence to *T. gondii* in these animals significantly increased with age as observed in cats from Spain (Gauss et al., 2003, Miró et al., 2004). *T. gondii* seroprevalences were similar in free-ranging (66.7% of 93) and wild-caught captive lynx (69% of 84) but significantly lower in captive-born lynx (22.5% of 40). Seroprevalence was higher in Iberian lynx with concurrent *Cytauxzoon felis* (88% of 25) but not with concurrent Feline Leukemia Virus (FeLV) infection (53.8% of 13). The study showed no significant differences in seroprevalence between sexes, geographic region and year of sample collection (García-Bocanegra et al., 2010b). The study examined the presence of oocysts of *T. gondii* but none were detected in fecal samples from 58 lynx analyzed (García-Bocanegra et al., 2010b). Rodríguez and Carbonell (1998) also failed to detect *T. gondii* in fecal samples from 9 lynx by coprological analysis.

The diet of free-ranging lynx mainly consists of wild rabbit (90.8%) with a smaller proportion of red-legged partridge (4.5%) and wild ruminants (2.8%) (Gil-Sánchez et al., 2006). Antibodies to *T. gondii* were found in 11.9% of rabbits tested from areas shared with the lynx (García-Bocanegra et al., 2010b). This prevalence was similar to those previously reported in southern Spain by Almería et al. (2004) and suggests that rabbit may be an important source of *T. gondii* infection for Iberian lynx. In addition, Iberian lynx could have access to carcasses from wild ruminant species, in which *T. gondii* antibodies have been reported in Sierra Morena (Gauss et al., 2006). Although seroprevalence of *T. gondii* in free-living red-legged partridges has not been reported, seroprevalence of *T. gondii* antibodies in several free-living avian species from South Spain have been reported recently (Cabezón et al., 2011).

The several studies performed in Iberian lynx indicate widespread *T. gondii* infection in the two areas where Iberian lynx survive in Spain. The fact that four captive-born lynx seroconverted (Garcia-Bocanegra et al., 2010b) was indication of contact with *T. gondii* in the Captive Breeding Centers, hence, control measures to prevent *T. gondii* infection also need to be implemented in these centers.

The high seroprevalence observed in the Iberian lynx suggests an important role in the sylvatic cycle of *T. gondii* in areas where the lynx remain.

2.3.2. European Wildcat (Felis silvestris silvestris)

There are only two published studies about *T. gondii* in European wildcats, both in central Spain. Rodríguez and Carbonell (1998) detected oocysts in faecal samples of 2 of 12 wildcats, while Sobrino et al., (2007) detected *T. gondii* antibodies in 3 of 6 European wildcats. Few studies have been performed on this species in other countries. In Great Britain, *T. gondii* antibodies were detected in 100% of 23 wildcats tested (Mc Orist, 1992) and Lukesova and Literak (1998) identified *T. gondii* oocysts in three wildcats living in a zoo in the Czech Republic.

3. Seroprevalence Studies of *Toxoplasma gondii* in Non-Feline Carnivores from Spain

3.1. Domestic Dog (*Canis familiaris*)

Toxoplasma gondii infection has been evaluated in dogs (*Canis familiaris*) in two studies in Spain (Millán et al., 2009b; Cabezón et al., 2010a). The first study detected a seroprevalence of antibodies against *T. gondii* in 53% of 24 dogs allowed to roam free from Andalusia, Southern Spain. These dogs lived outdoors and probably had access to intermediary hosts, such as rodents, rabbits or birds. The second serosurvey, performed in dogs from a kennel of Mallorca (Cabezón et al., 2010a) reported a higher seroprevalence (58.7% of 46 dogs analyzed) using the MAT (titer≥1:25), which was consistent with the high seroprevalence observed in feral cats in this Mediterranean island (Millán et al. 2009a). Cabezón et al. (2010a) reported that antibodies to *T. gondii* were more frequently observed in dogs older than 1 year (72%) than dogs younger than one year (28%). No other significant difference in exposure to the parasite was detected based on the age, sex, or the breed of the dogs. The worldwide seroprevalence of *T. gondii* in dogs was summarized by Dubey (2009a). Smielewska-Los et al. (2002) observed that dogs fed raw meat were significantly more frequently seropositive than those eating only commercial dry feed or cooked meat. Kennel dogs are more likely to have fed on rodents and/or rubbish when they are free roaming and thus more exposed to the pathogen. The higher seroprevalence in older dogs detected in Mallorca was in agreement with the previous results in cats, and possibly was due to a greater probability of exposure over time and long persistence of antibodies. These results confirm the widespread presence of *T. gondii* in dogs from the Mediterranean areas of Spain.

3.2. Non-Feline Wild Carnivores

The epidemiological role of non-felid wild and domestic carnivore species in *T. gondii* infection could be related to their use as sentinel species for these infections because they cannot act as definitive hosts since they are not capable of elimination of oocyts. However, some small species can be predated by felines, acting as intermediate host.

3.2.1. Non-Feline Wild Carnivore Mammals

With the exception of red foxes, there are few surveys of seroprevalence of *T. gondii* in non-feline wild carnivores worldwide. Seroprevalence (46.9%, 15 of 32, MAT) in wolves (*Canis lupus*) was analyzed by Sobrino et al., (2007) in Spain. This seroprevalence was higher than that observed in free-living wolves in Alaska, USA (9% of 125, MAT) (Zarnke et al., 2000) or in Canada (0% of 9, ELISA) (Philippa et al., 2004).

Other wild carnivore species were also included in the study of Sobrino et al., (2007). Antibodies to *T. gondii* were detected in 66 of 102 (64.7%) red foxes, 26 of 37 (70.3%) Eurasian badgers, 17 of 20 (85.0%) stone martens, 4 of 4 pine martens, 6 of 6 Eurasian otters, 4 of 4 polecats, 1 of 1 ferret, 13 of 21 (61.9%) European genets, and 13 of 22 (59.1%) Egyptian mongooses (see Table 1).

Overall seroprevalence of *T. gondii* antibodies in wild carnivores is higher when compared with non-carnivorous wild species such as rabbits, wild boar and wild ruminants (Almería et al., 2004; Gauss et al., 2005; Gauss et al.,

2006; García-Bocanegra et al., 2012a, respectively) or domestic species such as pigs, cattle and sheep or equids (García-Bocanegra et al., 2010a; Panadero et al., 2010; García-Bocanegra et al., 2012b, respectively). A higher seroprevalence of *T. gondii* infection in carnivore species is in accordance with the cycle of the parasite, where carnivores have greater exposure to infection when feeding on parasitized intermediary hosts (Smith and Frenkel; 1995).

In agreement with the findings reported in Spain (Sobrino et al., 2007), different surveys have shown a high seroprevalence to *T. gondii* in red foxes worldwide (Dubey et al., 1999; Murphy et al., 2007; Jakubek et al., 2007). In the study of Sobrino et al., (2007) high seroprevalences were also observed in several species of the Order Mustelidae (Table 1) and included the first report on the presence of antibodies in the Egyptian mongoose (Order Herpestidae) and in the European genet (Order Viverridae). *T. gondii* has been detected or isolated from these species worldwide (Thornton and Cook, 1986; Hejlícek et al., 1997; Hurková and Modrý, 2006; Anwar et al., 2006). Interestingly, Sobrino et al., (2007) reported, also for the first time in Spain, the presence of *T. gondii* antibodies in Eurasian otters (*Lutra lutra*). Recent studies have demonstrated water-borne transmission and wide dissemination of *T. gondii* in marine mammal populations, including sea otters, cetaceans and seals (Miller et al., 2002; Cabezón et al., 2004; Cabezón et al., 2010b, respectively). In this regard, river water contamination could be the main reason for the observed *T. gondii* infection of Eurasian otters in Spain.

3.2.2. Carnivore/Scavenger Wild Birds

Toxoplasma gondii infection has been evaluated in birds worldwide and has been detected in many avian species including poultry, game and other wild species (Dubey 2002; Dubey et al., 2010; Lopes et al., 2010; Molina-López et al., 2012; Darwich et al., 2012). Birds are considered as good sentinel species for the presence of *T. gondii* in the environment and for the presence of the parasite in other intermediary hosts (Gondim et al., 2010; Dubey and Jones, 2008). Cabezón et al. (2011) evaluated antibodies to *T. gondii* from sera obtained from 1079 wild birds belonging to 56 different species (including 610 Falconiformes, 260 Strigiformes, 156 Ciconiiformes, 21 Gruiformes, and 32 birds from other orders), from different areas of Spain. This study constituted the first extensive survey in wild bird species in Spain to date. Antibodies to *T. gondii* were found in 282 (26.1%) of the birds. The highest seropositivity by species was observed in the Eurasian eagle owl (*Bubo bubo*) (68.1%, 98 of 144). The main risk factors associated with *T. gondii* seropositivity in wild birds were age and diet, with the highest exposure in older and carnivorous wild birds (Cabezón et al., 2011). The results showed that *T. gondii* infection is widespread and can be at a high level in many wild birds in Spain, most likely related to their feeding behaviour. In addition, a high seroprevalence of *T. gondii* in common ravens (*Corvus corax*) (80.5% of 113 sera samples) was reported recently in Spain (Molina-López et al., 2012) and the presence of *T. gondii* DNA in brain tissues was detected in magpies (*Pica pica*), Eurasian jays (*Garrulus glandarius*), black kite and griffon vulture (*Gyps fulvus*) (Darwich et al., 2012), extending the list of natural intermediate hosts for *T. gondii* infection to these species.

4. Presence of Cats as Risk Factor on *T. gondii* Seroprevalence in Domestic Species in Spain

The presence of cats has been reported as one of the main factors for *T. gondii* seroprevalence in livestock species, such as small ruminants and pigs in Spain (Mainar et al., 1996; García-Bocanegra et al., 2010a, respectively). Cats are implicated in the maintenance of *T. gondii* infection in pig farms through oocyst elimination and contamination of feed and/or water (Dubey et al., 1995, Weigel et al., 1995). In this regard, the prevalence of *T. gondii* in pigs and mice decreased significantly in pig farms where cats were vaccinated with a modified live vaccine to reduce oocyst shedding by cats (Mateus-Pinilla et al., 1999). In addition, the presence of cats has also been also implicated in abortion storms in sheep in Spain (Mainar et al., 1996).

4.1. Domestic Pig

The consumption of raw or undercooked pork meat containing tissue cysts is considered one of the principal sources of human infection (Anonymous 2007, Dubey 2009b) and, in one instance acute toxoplasmosis, including blindness was reported in hunters whom had consumed uncooked pork from wild pigs (Choi et al., 1997).

In a recent study (García-Bocanegra et al., 2010a), serum samples from 2970 (1400 sows, 1570 fattening) pigs, from 100 farms in the 10 main swine production regions in Spain were tested for antibodies against *T. gondii* by MAT. Antibodies to *T. gondii* were detected in 492 pigs (16.6%, 9.7% in fattening pigs and 24.2% in sows). The presence of cats on the pig farms was a risk factor for *T. gondii* seroprevalence, in agreement with previous observations (Weigel et al., 1995; Assadi-Rad et al., 1995; Lehmann et al., 2003).

The seroprevalence observed in the study by García-Bocanegra et al., (2010a) indicates a widespread, although variable, exposure to T. gondii among domestic pigs in Spain, which may have important implications for public health. Management measures including control of cats and rodents on the farms could help to reduce the infection levels in Spain (García-Bocanegra et al., 2010a).

4.2. Small Ruminants

A seroepidemiological survey of *T. gondii* infection in sheep and goats conducted in the Madrid region of Spain in sera collected from 60 herds showed that seroprevalence was 11.8% (64 out of 541) (Mainar et al., 1996), (Table 1). Four variables were found to be significantly associated with seroprevalence assessed by multivariate analysis. Factors known to be linked with toxoplasmosis were the presence of cats, together with a previous history of abortion outbreaks on the farm. Another serosurvey of *T. gondii* in ruminants has been recently reported by Panadero et al. (2010) in samples from 177 sheep, 178 cattle and 160 roe deer (*Capreolus capreolus*) from pastures in Galicia (northwest Spain). The prevalence of antibodies to *T. gondii* was 13.7% in roe deer, 57% in sheep and 7.3% in cattle. Although it has not been evaluated in free-grazing ruminants, the most likely source of *T. gondii* infection for these animals would be free-living felids, namely feral cats and European wildcats, species known to have high rates of infection (Sobrino et al., 2007; Millán et al., 2009a; Millán et al., 2009b; García-Bocanegra et al., 2010b). In addition, Moreno et al., (2012) has recently investigated the presence of *T. gondii* in abortions in small ruminants detecting the parasite in 5.4% of 74 and 11.5% of 26 of ovine and caprine aborted fetus, respectively. Infection of ruminants is of public health concern.

4.3. Equids

García-Bocanegra et al. (2012b) recently reported for the first time the seroprevalence to *T. gondii* in equids from Spain and in donkeys in Europe. The results showed that *T. gondii* antibodies are moderately widespread in herds of equine species in southern Spain. Around 18% of 420 herd horses showed the presence of antibodies with an individual seroprevalence in horses of 10.8% (49/454) (García-Bocanegra et al., 2012b) which is similar to that reported in some other European countries such as Czech Republic (8%) (Hejlíček and Literák, 1994), Turkey (7.2-8%) (Zeybek et al., 1998; Karatepe et al., 2010) and the Netherlands (7%) (van Knapen et al., 1982).

The seroprevalence observed in donkeys (25.6%) was significantly higher than in horses. This could be due to the fact that donkeys are kept more frequently as free ranging animals in the field, where they could have more contact with cats around stables and corrals as indicated by El-Ghaysh (1998). Equids can only become infected by ingestion of food and water contaminated with sporulated *T. gondii* oocysts from felids (Dubey and Beattie, 1988, Dubey, 2009) or by congenital transmission, which has also been reported in horses (Turner and Savva, 1992). In the study by García-Bocanegra et al. (2012b) although higher seropositivity was observed in equine herds with the presence of cats, the differences were not statistically significant.

T. gondii has been isolated from experimentally infected ponies (Al-Khalidi et al., 1980) and more recently in 52.6% of 150 horse meat samples from naturally apparently healthy horses in Egypt (Shaapan and Ghazy, 2007). Consumption of horse meat could therefore be a potential source of human infection. A study showed that *T. gondii* can persist in edible tissues of living equines for up to 476 days and so, it has been suggested that non-cooked horse meat should not be fed to cats or eaten by human beings (Dubey, 1985). Although horses in Spain are not primarily kept for slaughter, a small proportion of meat is used for human consumption and, occasionally horse meat is used to feed felid species in zoos. Therefore, equines could have a role in the transmission of *T. gondii* to humans and to cats in Spain.

5. Aquatic Environment

Until recently, waterborne transmission of *T. gondii* was considered uncommon. However, a large human outbreak linked to contamination of a municipal water reservoir in Canada by wild felids and the widespread infection of marine mammals in the USA provided two reasons to question this view (Jones and Dubey, 2010).

Toxoplasma gondii infection in marine mammals can serve as indication of contamination of marine waters with oocysts, and is therefore considered a good sentinel for protozoan contamination of marine waters (Van Bressen et al., 2009). In the Spanish Mediterranean area, Domingo et al., (1992) and

Resendes et al., (2002) described fatal disease in Striped dolphins and disease with transplacental infection in Risso's dolphin. Moreover, Cabezón et al., (2004) reported the presence of *T. gondii* antibodies in 4 of 36 striped dolphins, in 2 of 4 common dolphins (*Delphinus delphis*), in 4 of 7 bottlenose dolphins (*Tursiops truncatus*), and in 1 of 1 harbour porpoise (*Phocoena phocoena*). Antibodies were not found in 9 Risso's dolphins and in 1 long-finned pilot whale (*Globicephala melas*) (Cabezón et al., 2004). These results indicated that *T. gondii* infection is frequent in at least 3 dolphin species from the Mediterranean Sea and indicated that this environment is contaminated by *T. gondii* oocysts.

Interestingly, some piscivorous wild birds studied by Cabezón et al., (2011) showed contact with the parasite, including species such as the Eurasian spoonbill (*Platelea leucorodia*), of a large sample size (6.2% of 81 samples). This result also could be an indication of continental water contamination by oocysts.

6. *Toxoplasma gondii* in Pregnant Women in Spain

According to de Ory Manchon et al., (2009) the prevalence of *T. gondii* antibodies has decreased in pregnant women in Spain during the last 20 years. Nevertheless, seroprevalence of *T. gondii* in pregnant women in Spain have been reported to range from 18.8% in 2,929 pregnant women in Salamanca (Gutiérrez-Zufiaurre et al., 2004) to as high as 50% in pregnant women from Catalonia (Pumarola et al., 1989; Pujol-Piqué et al., 2000; Muñoz-Batet et al., 2004).

Based on these data and our previous studies in animal species, Catalonia seems to be one of the most prevalent areas of *T. gondii* infection in Spain compared to other locations for species such as wild rabbits (Almería et al., 2004), roe deer (Gamarra et al., 2008), domestic pigs (García-Bocanegra et al., 2010a) and wild birds (Cabezón et al., 2011); and where higher detection of the parasite has also been observed in tissues from wild birds (Darwich et al., 2012). The moderate precipitation and humidity in Catalonia might favour oocyst survival and sporulation in the environment facilitating *T. gondii* spread and maintenance (Dubey and Beattie, 1988). Furthermore, there are abundant populations of free-roaming cats, especially in the urban areas around Barcelona (Gauss et al., 2003).

To the best of our knowledge, only a single study has characterized the genotypes of *T. gondii* in human specimens in Spain: Fuentes et al. (2001) showed that Type-I were mainly found in congenital infection samples, while Type-II were mainly found in immunocompromised patients in the Madrid area (central Spain). The genotypes of *T. gondii* found in humans concur with the types characterized in cats sampled (Montoya et al., 2008). However, the two studies above are limited both in terms of geography and in the variety of studied species. Therefore, further characterization of the specific genotypes present in Spain in humans and animals in other areas is needed.

Conclusion

Cats have a prominent role in the transmission of *T. gondii* infection. Recent studies in different felid species, including both domestic and wild felids from different areas of Spain, have shown elevated seroprevalence levels. Furthermore, the presence of cats has been reported as one of the main factors for *T. gondii* seroprevalence in other domestic species, such as pigs and small ruminants in Spain.

Since seropositive felids are likely to have already shed *T. gondii* oocysts in the environment, the high rate of *T. gondii* seroprevalence in cats can have has important implications for conservation and animal infection and public health in Spain. To date, not many few studies have been carried out to analyze the genotypes present in cats, humans and other animal species in Spain.

In this regard, further characterization of the specific genotypes present in humans and in animals, especially in areas such as Catalonia and particularly Majorca, which seems to be one of the most prevalent areas of *T. gondii* infection in Spain, are needed to establish adequate prophylactic and control measures against this infection.

Ackowledgements

This study received partial support from the Spanish CICYT grants AGL2007-65521 C02/GAN and AGL2010-21273-C03/GAN. We will like to thank everyone involved in the many epidemiological studies performed by our research group, particularly to Dr. J.P. Dubey for his constant support.

REFERENCES

Al-Khalidi NW, Weisbrode SE, Dubey JP (1980) Pathogenicity of *Toxoplasma gondii* oocysts to ponies. *Am. J. Vet. Res.* 41, 1549-1551.

Almería S, Calvete C, Pagés A, Gauss C, Dubey JP (2004) Factors affecting the seroprevalence of *Toxoplasma gondii* infection in wild rabbits (*Oryctolagus cuniculus*) from Spain. *Vet. Parasitol.* 123, 265-270.

Anonymous (2007) Scientific opinion of the Panel on Biological Hazards on a request from EFSA on surveillance and monitoring of *Toxoplasma* in humans, foods and animals. *EFSA* J 583, 1–64.

Anwar A, Knaggs J, Service KM, McLaren GW, Riordan P, Newman C, Delahay RJ, Cheesman C, Macdonald DW (2006) Antibodies to *Toxoplasma gondii* in Eurasian badgers. *J. Wildl. Dis.* 42, 179-81.

Aparicio Garrido J, Cour Boveda I, Berzosa Aguilar AM, Pareja Miralles J (1972) Study on the epidemiology of toxoplasmosis. Infection in the domestic cat in suburbs of Madrid. Serological and copro-parasitological study. *Med. Trop.* (Madr) 48, 24-39.

Assadi-Rad AM, New JC, Patton S (1995) Risk factors associated with transmission of *Toxoplasma gondii* to sows kept in different management systems in Tennessee. *Vet. Parasitol.* 57, 289–297.

Basso W, Edelhofer R, Zenker W, Möstl K, Kübber-Heiss A, Prosl H (2005) Toxoplasmosis in Pallas' cats (*Otocolobus manul*) raised in captivity. Parasitology 130, 293-239. Erratum in: Parasitology 130, 586.

Bermúdez R, Faílde LD, Losada AP, Nieto JM, Quiroga MI (2009) Toxoplasmosis in Bennett's wallabies (*Macropus rufogriseus*) in Spain. *Vet. Parasitol.* 160, 155-158.

Cabezón O, Resendes AR, Domingo M, Raga JA, Agustí C, Alegre F, Mons JL, Dubey JP, Almería S (2004) Seroprevalence of *Toxoplasma gondii* antibodies in wild dolphins from the Spanish Mediterranean coast. *J. Parasitol.* 90, 643-644.

Cabezón O, Millán J, Gomis M, Dubey JP, Ferroglio E, Almería S (2010a) Kennel dogs as sentinels of *Leishmania infantum*, *Toxoplasma gondii*, and *Neospora caninum* in Majorca Island, Spain. *Parasitol. Res.* 107, 1505-1508.

Cabezón O, Hall AJ, Vincent C, Pabón M, García-Bocanegra I, Dubey JP, Almería S (2010b) Seroprevalence of *Toxoplasma gondii* in North-eastern Atlantic harbor seal (*Phoca vitulina vitulina*) and grey seal (*Halichoerus grypus*). *Vet. Parasitol.* 179, 253-256.

Cabezón O, García-Bocanegra I, Molina-López R, Marco I, Blanco JM, Höfle U, Margalida A, Bach-Raich E, Darwich L, Echeverría I, Obón E, Hernández M, Lavín S, Dubey JP, Almería S. (2011) Seropositivity and risk factors associated with *Toxoplasma gondii* infection in wild birds from Spain. *PLoS One*. doi:10.1371/journal.pone.0029549.

Choi WY, Nam HW, Kwak NH, Huh W, Kim YR, Kang MW, Cho SY, Dubey JP (1997) Foodborne outbreaks of human toxoplasmosis. *J. Infect. Dis.* 175, 1280–1282.

Closa-Sebastià F, Casas-Díaz E, Cuenca R, Lavín S, Mentaberre G, Marco I (2011) Antibodies to selected pathogens in wild boar (*Sus scrofa*) from Catalonia (NE Spain). *Eur. J. Wildl. Res.* 57, 977–981.

Cook AJ, Gilbert RE, Buffolano W, Zufferey J, Petersen E, Jenum PA, Foulon W, Semprini AE, Dunn DT (2000) Sources of *Toxoplasma* infection in pregnant women: European multicentre case-control study. European Research Network on Congenital Toxoplasmosis. *BMJ* 321, 142-147.

Darwich L, Cabezón O, Echeverria I, Pabón M, Marco I, Molina R, Alarcia-Alejos O, López-Gatius F, Lavín S, Almería S (2012) Presence of *Toxoplasma gondii* and *Neospora caninum* DNA in brain of wild birds. *Vet. Parasitol.* 183, 77-81.

Domingo M, Visa J, Pumarola M, Marco AJ, Ferrer L, Rabanal R, Kennedy S (1992) Pathologic and immunocytochemical studies of morbillivirus infection in striped dolphins (*Stenella coeruleoalba*). *Vet. Pathol.* 29, 1-10.

Dubey JP (1973) Feline toxoplasmosis and coccidiosis: a survey of domiciled and stray cats. *J. Am. Vet. Med. Assoc.* 162, 873-877.

Dubey JP (1985) Persistence of encysted *Toxoplasma gondii* in tissues of equids fed oocysts. *Am. J. Vet. Res.* 46, 1753-4.

Dubey JP (2002) A review of Toxoplasmosis in wild birds. *Vet. Parasitol.* 106:121–153.

Dubey JP (2009a) Toxoplasmosis in animals and humans. Second Edition. Boca Raton, FL: CRC Press. p.1-313.

Dubey JP (2009b) Toxoplasmosis in pigs. The last 20 years. *Vet. Parasitol* 164, 89-103.

Dubey JP, Beattie CP (1988) *Toxoplasmosis of Animals and Man.* Boca Raton, FL: CRC Press. 1988. p. 1–220.

Dubey JP, Jones JL (2008) *Toxoplasma gondii* infection in humans and animals in the United States. *Int. J. Parasitol.* 38, 1257-1278.

Dubey JP, Hamir AN, Rupprecht CE (1990) Acute disseminated toxoplasmosis in a red fox (*Vulpes vulpes*). *J. Wildl. Dis.* 26,286-290.

Dubey JP, Weigel RM, Siegel AM, Thulliez P, Kitron UD, Mitchell MA, Mannelli A, Mateus-Pinilla NE, Shen SK, Kwok OCH, Todd KS (1995) Sources and reservoirs of *Toxoplasma gondii* infection on 47 swine farms in Illinois. *J. Parasitol* 81, 723–729.

Dubey JP, Speer CA, Shen SK, Kwok OCH, Blixt JA (1997) Oocyst-induced murine toxoplasmosis: Life cycle, pathogenicity, and stage conversion in mice fed *Toxoplasma gondii* oocysts. *J. Parasitol.* 83, 870–882.

Dubey JP, Storandt ST, Kwok OC, Thulliez P, Kazacos KR (1999) *Toxoplasma gondii* antibodies in naturally exposed wild coyotes, red foxes, and gray foxes and serologic diagnosis of Toxoplasmosis in red foxes fed *T. gondii* oocysts and tissue cysts. *J. Parasitol.* 85, 240-243.

Dubey JP, Saville W J A, Stanek JF, Reed SM (2002). Prevalence of *Toxoplasma gondii* antibodies in domestic cats from rural Ohio. *J. Parasitol.* 88, 802–803.

Dubey JP, Velmurugan GV, Ulrich V, Gill J, Carstensen M, Sundar N, Kwok OCH, Thulliez P, Majumdar D, Su C (2008) Transplacental toxoplasmosis in naturally-infected white-tailed deer: Isolation and genetic characterisation of *Toxoplasma gondii* from foetuses of different gestational ages. *Int. J. Parasitol.* 38, 1057-1063.

Dubey JP, Felix TA, Kwok OC (2010) Serological and parasitological prevalence of *Toxoplasma gondii* in wild birds from Colorado. *J. Parasitol.* 96, 937-939.

El-Ghaysh A (1998) Seroprevalence of *Toxoplasma gondii* in Egyptian donkeys using ELISA. *Vet. Parasitol.* 80, 71-73.

Fuentes I, Rubio JM, Ramírez C, Alvar J (2001) Genotypic characterization of *Toxoplasma gondii* strains associated with human toxoplasmosis in Spain: direct analysis from clinical samples. *J. Clin. Microbiol.* 39, 1566-1570.

Gamarra JA, Cabezón O, Pabón M, Arnal MC, Luco DF, Dubey JP, Gortázar C, Almeria S (2008) Prevalence of antibodies against *Toxoplasma gondii* in roe deer from Spain. *Vet. Parasitol.* 153, 152-156.

García-Bocanegra I, Simon-Grifé M, Dubey JP, Casal J, Martín GE, Cabezón O, Perea A, Almería S (2010a) Seroprevalence and risk factors associated with *Toxoplasma gondii* in domestic pigs from Spain. *Parasitol. Int.* 59, 421-426.

García-Bocanegra I, Dubey JP, Martínez F, Vargas A, Cabezón O, Zorrilla I, Arenas A, Almería S (2010b) Factors affecting seroprevalence of *Toxoplasma gondii* in the endangered Iberian lynx (*Lynx pardinus*). *Vet. Parasitol.* 167, 36-42.

García-Bocanegra I, Cabezón O, Pabón M, Gómez-Guillamón F, Arenas A, Alcaide E, Salas-Vega R, Dubey JP, Almería S (2012a) Prevalence of *Toxoplasma gondii* and *Neospora caninum* antibodies in Spanish ibex (*Capra pyrenaica hispanica*). *Vet. J.* 191, 257-260.

García-Bocanegra I, Cabezón O, Arenas-Montes A, Carbonero A, Dubey JP, Perea A, Almería S (2012b) Seroprevalence of *Toxoplasma gondii* in equines from Andalusia (Southern Spain). Parasitol Int (in press). http://dx.doi.org/10.1016/j.parint.2012.02.003.

Gauss CB, Almería S, Ortuño A, García F, Dubey JP (2003) Seroprevalence of *Toxoplasma gondii* antibodies in domestic cats from Barcelona, Spain. *J. Parasitol.* 89, 1067-1068.

Gauss CB, Dubey JP, Vidal D, Ruiz F, Vicente J, Marco I, Lavin S, Gortazar C, Almería S (2005) Seroprevalence of *Toxoplasma gondii* in wild pigs (*Sus scrofa*) from Spain. *Vet. Parasitol.* 131, 151-156.

Gauss CB, Dubey JP, Vidal D, Cabezón O, Ruiz-Fons F, Vicente J, Marco I, Lavin S, Gortazar C, Almería S (2006) Prevalence of *Toxoplasma gondii* antibodies in red deer (*Cervus elaphus*) and other wild ruminants from Spain. *Vet. Parasitol.* 136, 193-200.

Gil-Sánchez JM, Ballesteros-Duperon E, Bueno-Segura JF (2006) Feeding ecology of the Iberian lynx (*Lynx pardinus*) in eastern Sierra Morena (Southern Spain). *Acta Theriol.* 51, 85–90.

Gondim LS, Abe-Sandes K, Uzêda RS, Silva MS, Santos SL, Mota RA, Vilela SM, Gondim LF (2010) *Toxoplasma gondii* and *Neospora caninum* in sparrows (*Passer domesticus*) in the Northeast of Brazil. *Vet. Parasitol.* 168, 121-124.

Gutiérrez-Zufiaurre N, Sánchez-Hernández J, Muñoz S, Marín R, Delgado N, Sáenz MC, Muñoz-Bellido JL, García-Rodríguez JA (2004) Seroprevalence of antibodies against *Treponema pallidum*, *Toxoplasma gondii*, rubella virus, hepatitis B and C virus, and HIV in pregnant women. *Enferm. Infecc. Microbiol. Clin.* 22, 512-516.

Hejlíček, K., Literák, I (1994) Prevalence of antibodies to *Toxoplasma gondii* in horses in the Czech Republic. *Acta Parasitol.* 39, 217–219.

Hejlícek K, Literák I, Nezval J (1997) Toxoplasmosis in wild mammals from the Czech Republic. *J. Wildl. Dis.* 33, 480-485.

Hurková L, and Modrý D (2006) PCR detection of *Neospora caninum*, *Toxoplasma gondii* and *Encephalitozoon cuniculi* in brains of wild carnivores. *Vet. Parasitol.* 137, 150-154.

IUCN - International Union for Conservation of Nature and Natural Resources. –Red List 2011, www.iucnredlist.org.

Jakubek EB, Lundén A, Uggla A (2006) Seroprevalences of *Toxoplasma gondii* and *Neospora* sp. infections in Swedish horses. *Vet. Parasitol.* 138, 194-199.

Johnson AM, Roberts H, Statham P, Munday BL (1989) Serodiagnosis of acute toxoplasmosis in macropods. *Vet. Parasitol.* 34, 25-33.

Jones, J. L., Sehgal, M., and Maguire, J. H. (2002) Toxoplasmosis-associated deaths among human immunodeficiency virus-infected persons in the United States, 1992–1998. *Clin. Infect. Dis.* 34, 1161.

Jones JL, Dubey JP (2010) Waterborne toxoplasmosis--recent developments. *Exp. Parasitol.* 124: 10-25.

Juan-Sallés C, Prats N, López S, Domingo M, Marco AJ, Morán JF (1997) Epizootic disseminated toxoplasmosis in captive slender-tailed meerkats (*Suricata suricatta*). *Vet. Pathol.* 34, 1-7.

Juan-Sallés C, Prats N, Marco AJ, Ramos-Vara JA, Borrás D, Fernández J (1998) Fatal acute toxoplasmosis in three golden lion tamarins (*Leontopithecus rosalia*). *J. Zoo Wildl. Med.* 29, 55-60.

Juan-Sallés C, Mainez M, Marco A, Sanchís AM (2011) Localized toxoplasmosis in a ring-tailed lemur (*Lemur catta*) causing placentitis, stillbirths, and disseminated fetal infection. *J. Vet. Diagn. Invest.* 23, 1041-1045.

Karatepe B, Babür C, Karatepe M, Kiliç S (2010) Seroprevalence of toxoplasmosis in horses in Niğde Province of Turkey. *Trop. Anim. Health Prod.* 42, 385-389.

Lehmann T, Graham DH, Dahl E, Sreekumar C, Launer F, Corn JL, Gamble HR, Dubey JP (2003) Transmission dynamics of *Toxoplasma gondii* on a pig farm. *Infect. Genet. Evol.* 3, 135–141.

Liberg O, Sandell M, Pontier D, Natoli E (2000) Density, spatial organization and reproductive tactics in the domestic cat and other felids. In: Turner, D.C., Bateson, P. (Eds.), The Domestic Cat. The Biology of its Behaviour. Cambridge University Press, Cambridge, pp. 119–148.

Lloyd C, Stidworthy MF (2007) Acute disseminated toxoplasmosis in a juvenile cheetah (*Acinonyx jubatus*). *J. Zoo Wildl. Med.* 38, 475-478.

Lopes AP, Sargo R, Rodrigues M, Cardoso L (2010) High seroprevalence of antibodies to *Toxoplasma gondii* in wild animals from Portugal. *Parasitol. Res.* 108, 1163-1169.

Lukesová D, Literák I (1998) Shedding of *Toxoplasma gondii* oocysts by Felidae in zoos in the Czech Republic. *Vet. Parasitol.* 74, 1-7.

Mainar RC, de la Cruz C, Asensio A, Domínguez L, Vázquez-Boland JA (1996) Prevalence of agglutinating antibodies to *Toxoplasma gondii* in small ruminants of the Madrid region, Spain, and identification of factors influencing seropositivity by multivariate analysis. *Vet. Res. Commun.* 20, 153-159.

Marco I, Velarde R, López-Olvera JR, Cabezón O, Pumarola M, Lavín S (2009) Systemic toxoplasmosis and Gram-negative sepsis in a southern chamois (*Rupicapra pyrenaica*) from the Pyrenees in northeast Spain. *J. Vet. Diagn. Invest.* 21, 244-247.

Mateus-Pinilla NE, Dubey JP, Choromanski L, Weigel RM (1999) A field trial of the effectiveness of a feline *Toxoplasma gondii* vaccine in reducing *T. gondii* exposure for swine. *J. Parasitol.* 85, 855–860.

Mc Orist S (1992) Diseases of the European wildcat (Felis silvestris Schreber, 1777) in Great Britain. *Rev. Sci. Tech.* 11, 1143-1149.

Millán J, Candela MG, Palomares F, Cubero MJ, Rodríguez A, Barral M, de la Fuente J, Almería S, León-Vizcaíno L (2009a) Disease threats to the endangered Iberian lynx (*Lynx pardinus*). *Vet. J.* 182, 114-124.

Millán J; Cabezón O; Pabón M; Dubey JP; Almería S (2009b) Seroprevalence of *Toxoplasma gondii* and *Neospora caninum* in feral cats (*Felis silvestris catus*) in Majorca, Balearic Islands, Spain. *Veterinary Parasitology*, 165, 323-326.

Millán J (2010) Feeding habits of feral cats *Felis silvestris catus* in the countryside of Majorca Island, Spain. *Wildlife Biology in Practice* 6:32-38.

Miller MA, Gardner IA, Kreuder C, Paradies DM, Worcester KR, Jessup DA, Dodd E, Harris MD, Ames JA, Packham AE, Conrad PA. (2002) Coastal freshwater runoff is a risk factor for *Toxoplasma gondii* infection of southern sea otters (*Enhydra lutris nereis*). *Int. J. Parasitol.* 32, 997-1006.

Miró G, Montoya A, Jiménez S, Frisuelos C, Mateo M, Fuentes I (2004) Prevalence of antibodies to *Toxoplasma gondii* and intestinal parasites in stray, farm and household cats in Spain. *Vet. Parasitol.* 126, 249-55.

Molina-López R, Cabezón O, Pabón M, Darwich L, Obón E, Lopez-Gatius F, Dubey JP, Almería S (2012) High seroprevalence of *Toxoplasma gondii* and *Neospora* caninum in the Common raven (*Corvus corax*) in the Northeast of Spain. Res Vet Sci http://dx.doi.org/10.1016/j.rvsc.2011.05.011.

Montoya A, Miró G, Mateo M, Ramírez C, Fuentes I (2008) Molecular characterization of *Toxoplasma gondii* isolates from cats in Spain. *J. Parasitol.* 94, 1044-1446.

Moreno B, Collantes-Fernández E, Villa A, Navarro A, Regidor-Cerrillo J, Ortega-Mora LM (2012) Occurrence of *Neospora caninum* and *Toxoplasma gondii* infections in ovine and caprine abortions. Vet Parasitol. In press. http://dx.doi.org/10.1016/j.vetpar.2011.12.034.

Muñoz-Batet C, Guardià-Llobet C, Juncosa-Morros T, Viñas-Domenech L, Sierra-Soler M, Sanfeliu-Sala I, Bosch-Mestres J, Dopico-Ponte E, Lite-Lite J, Matas-Andreu L, Juste-Sánchez C, Barranco-Romeu M (2004) Toxoplasmosis and pregnancy. Multicenter study of 16,362 pregnant women in Barcelona. *Med. Clin.* (Barc) 123, 12-16.

Murphy TM, Walochnik J, Hassl A, Moriarty J, Mooney J, Toolan D, Sanchez-Miguel C, O'Loughlin A, McAuliffe A (2007) Study on the prevalence of *Toxoplasma gondii* and *Neospora caninum* and molecular evidence of *Encephalitozoon cuniculi* and *Encephalitozoon* (Septata) *intestinalis* infections in red foxes (*Vulpes vulpes*) in rural Ireland. *Vet. Parasitol.* 146, 227-234.

Nicolle C and Manceaux L (1909) Sur un protozoaire nouveau du gondi. Archives de l'Institut Pasteur 2, 97–103.

Nowell K and Jackson P (1996) *Wild Cats. Status Survey and Conservation Action Plan.* IUCN/SSC Cat Specialist Group, Gland, Switzerland and Cambridge, UK.

Ory Manchón F (2009) Seroepidemiological surveys of non vaccine-preventable diseases and their interest in public health. *Rev. Esp. Salud. Pública* 83, 645-57.

Palomares F, Godoy JA, Piriz A, O'Brien SJ, Johnson WE (2002) Faecal genetic analysis to determine the presence and distribution of elusive carnivores: design and feasibility for the Iberian lynx. *Molec. Ecol.* 11, 2171-2182.

Palomo LJ, Gisbert J (2002) Atlas de los mamíferos terrestres de España. Dirección General de Conservación de la Naturaleza. SECEM-SECEMU, Madrid, Spain.

Panadero R, Painceira A, López C, Vázquez L, Paz A, Díaz P, Dacal V, Cienfuegos S, Fernández G, Lago N, Díez-Baños P, Morrondo P (2010) Seroprevalence of *Toxoplasma gondii* and *Neospora caninum* in wild and domestic ruminants sharing pastures in Galicia (Northwest Spain). *Res. Vet. Sci.* 88, 111-115.

Pereira-Bueno J, Quintanilla-Gozalo A, Pérez-Pérez V, Alvarez-García G, Collantes-Fernández E, Ortega-Mora LM (2004) Evaluation of ovine abortion associated with *Toxoplasma gondii* in Spain by different diagnostic techniques. Vet Parasitol 121:33-43. Erratum in: *Vet. Parasitol.* (2004) 121:353.

Philippa JD, Leighton FA, Daoust PY, Nielsen O, Pagliarulo M, Schwantje H, Shury T, Van Herwijnen R, Martina BE, Kuiken T, Van de Bildt MW, Osterhaus AD (2004) Antibodies to selected pathogens in free-ranging terrestrial carnivores and marine mammals in Canada. *Vet. Rec.* 155, 135–140.

Pujol-Riqué M, Quintó L, Danés C, Valls ME, Coll O, Moreno A, Jiménez de Anta MT (2000) Dating anti-*Toxoplasma* IgM in pregnancy using VIDAS-ELFA methods. *Enferm. Infecc. Microbiol. Clin.* 18, 274-278.

Pumarola A, Salleras L, Vidal J, Canela J, Mas J, Jiménez de Anta MT, Pumarola T, Coll JJ, de la Puente ML (1989) Prevalence of antibodies against *Toxoplasma gondii*, rubella virus, cytomegalovirus and herpes simplex virus in pregnant women of Catalonia. *Enferm. Infecc. Microbiol. Clin.* 7, 83-86.

Remington JS, McLeod R, Thulliez P, and Desmonts G (2006) Toxoplasmosis. In ''Infectious Disease of the Fetus and Newborn Infant, 6th edition'' (Remington, Klein, Wilson, and Baker, eds.), pp. 947–1091. Elsevier Saunders, Philadelphia.

Resendes AR, Almería S, Dubey JP, Obón E, Juan-Sallés C, Degollada E, Alegre F, Cabezón O, Pont S, Domingo M (2002). Disseminated toxoplasmosis in a Mediterranean pregnant Risso's dolphin (*Grampus griseus*) with transplacental fetal infection. *J. Parasitol.* 88:1029-1032.

Riera M, Altés J, Villajonga C, Salas A, Homar F, Picco G, Leyes M, Rifà J (1995) Cambios en la historia natural del sida en pacientes de Mallorca e Ibiza: 1986–1992. *Med. Clin.* (Barcelona) 104, 765–770.

Rodríguez A, Carbonell E (1998) Gastrointestinal parasites of the Iberian lynx and other wild carnivores from central Spain. *Acta Parasitologica* 43, 128-136.

Roelke ME, Johnson WE, Millán J, Palomares F, Revilla E, Rodríguez A, Calzada J, Ferreras P, León-Vizcaíno L, Delibes M, O'Brien SJ (2008) Exposure to disease agents in the endangered Iberian lynx (*Lynx pardinus*). *Eur. J. Wildl. Res.* 54, 171–178.

Shaapan RM, Ghazy AA (2007) Isolation of *Toxoplasma gondii* from horse meat in Egypt. *Pak. J. Biol. Sci.* 10, 174-177.

Smielewska-Los E, Rypuła K, Pacoń J (2002) The influence of feeding and maintenance system on occurrence of *Toxoplasma gondii* infections in dogs. *Pol. J. Vet. Sci.* 5, 231-235.

Smith DD, Frenkel JK (1995) Prevalence of antibodies to *Toxoplasma gondii* in wild mammals of Missouri and east central Kansas: biologic and ecologic considerations of transmission. *J. Wildl. Dis.* 31, 15–21.

Sobrino R, Cabezón O, Millán J, Pabón M, Arnal MC, Luco DF, Gortázar C, Dubey JP, Almeria S (2007) Seroprevalence of *Toxoplasma gondii* antibodies in wild carnivores from Spain. *Vet. Parasitol.* 148:187-192.

Sunquist Man, Sunquist F (2002) Wild Cats of the World. University of Chicago Press.

Thornton RN, Cook TG (1986) A congenital *Toxoplasma*-like disease in ferrets (*Mustela putorius furo*). *N. Z. Vet. J.* 34, 31-33.

Turner CB, Savva D (1990) Evidence of *Toxoplasma gondii* in an equine placenta. *Vet. Rec.* 127, 96.

Van Bressem MF, Raga JA, Di Guardo G, Jepson PD, Duignan PJ, Siebert U, Barrett T, Santos MC, Moreno IB, Siciliano S, Aguilar A, Van Waerebeek K (2009) Emerging infectious diseases in cetaceans worldwide and the possible role of environmental stressors. *Dis. Aquat. Organ.* 86, 143-157. Review.

Van Knapen F, Franchimont JH, van der Lugt G (1982) Prevalence of antibodies to *Toxoplasma gondii* in farm animals in The Netherlands and its implication for meat inspection. *Vet. Quarterly* 4, 101-105.

Weigel RM, Dubey JP, Siegel AM, Kitron UD, Mannelli A, Mitchell MA, Mateus-Pinilla NE, Thulliez P, Shen SK, Kwok OCH, Todd KS (1995) Risk factors for transmission of *Toxoplasma gondii* on swine farms in Illinois. *J. Parasitol.* 81, 736–741.

Zarnke RL, Dubey JP, Kwok OCH, Ver Hoef JM (2000) Serologic survey for *Toxoplasma gondii* in selected wildlife species from Alaska. *J. Wildl. Dis.* 36, 219–224.

Zeybek H, Dundar B, Altintas K, Gungor C (1998) Tek tirnakli hayvanlarda toxoplasmosis'in seroprevalansi. *Acta Parasitol. Turc.* 22, 424-427.

In: Felines
Editors: R. García and R. Jiménez

ISBN 978-1-62257-005-8

Chapter 5

FELINE UVEITIS: THERAPEUTIC EFFECT OF MELATONIN

Ruth E. Rosenstein*, Pablo H. Sande, Diego C. Fernandez, Marcelo A. Aba[1], María Inés Keller Sarmiento, Daniel A. Sáenz, and María José Del Sole[1]

Laboratory of Retinal Neurochemistry and Experimental Ophthalmology, Department of Human Biochemistry, School of Medicine, University of Buenos Aires, CEFyBO/CONICET, Buenos Aires, Argentina

[1]Laboratory of Nervous System Physiology and Endocrinology, Department of Physiopathology, School of Veterinary Science, National University of the Center of Buenos Aires Province, Tandil, Buenos Aires Province, Argentina

ABSTRACT

Acute or chronic uveitis is a common ophthalmic disorder that constitutes one of the main causes of irreversible blindness in domestic cats. Uveitis affects primarily the uveal tract, but other ocular critical

*Corresponding Author: Dr. Ruth E. Rosenstein. Department of Human Biochemistry, School of Medicine, CEFyBO, University of Buenos Aires, CONICET, Paraguay 2155, 5th floor, (1121). Buenos Aires, ARGENTINA. Tel: 54-11-45083672 (ext 37). Fax: 54-11-45083672 (ext 317). e-mail: ruthr@fmed.uba.ar.

structures can also be affected. In addition, inflammation of the posterior segment (choroid and retina) of the eye may provoke retinal detachment and degeneration, with consequent blindness. At present, treatment of cats with uveitis generally involves administration of corticosteroids, which is directed toward reduction of pain and lesions in ocular tissues. However, the immunosuppressive effect of corticosteroids may contribute to the development of the systemic disease, and their chronic use could result in corticosteroid-induced glaucoma. A novel method for experimental induction of uveitis in cats would greatly facilitate the understanding of the cellular events involved in ocular inflammation as well as the development of new treatment strategies. In this chapter, we will summarize recent investigations showing that a single intravitreal injection of bacterial lipopolysaccharide (LPS) in cat eyes induces clinical, biochemical, functional, and histological changes that are consistent with the main features of naturally occurring feline uveitis, and that melatonin prevents clinical, biochemical, histological, ultrastructural, and functional consequences of experimental uveitis in cats.

Keywords: Melatonin, uveitis, nitric oxide, oxidative damage, feline uveitis

ABBREVIATIONS

AA: arachidonic acid;
AA-NAT: arylalkylamine N-acetyltransferase;
AH: aqueous humor;
BOB: blood-ocular barrier;
BAB: blood-aqueous barrier;
BRB: blood-retinal barrier;
COX: cyclooxygenase;
EAU: experimental autoimmune uveitis;
EIU: endotoxin induced uveitis;
ERG: electroretinogram;
HIOMT: hydroxyl-indole-O-methyltransferase;
IL: interleukin;
iNOS: inducible isoform of NO synthase;
IOP: intraocular pressure;
LIU: lens-induced uveitis;
LPS: lipopolysaccharide;
MT1: melatonin receptor type 1;
MT2: melatonin receptor type 2;

NOS: NO synthase;
NSAIDs: nonsteroidal anti-inflammatory drugs;
PGs: prostaglandins;
ROS: reactive oxygen species;
RPE: retinal pigment epithelium;
TNF-α: tumor necrosis factor α.

FELINE UVEITIS

Uveitis is the term used to describe many forms of intraocular inflammation, involving the uveal tract (iris, ciliary body, and choroid) and adjacent ocular structures (retina, vitreous, and optic nerve). Depending on the structures involved in the inflammatory process, uveitis is classified in anterior, intermediate, and posterior uveitis. In anterior uveitis, the iris (iritis) and/or the anterior part of the ciliary body (iridocyclitis) are affected; intermediate uveitis affects the posterior part of the ciliary body (*parsplanitis*), while posterior uveitis primarily involves the choroid (choroiditis), although due to its proximity to the retina, the inflammation of both structures (retinochoroiditis or chorioretinitis) is usual (Wilkie, 1994; Collins and Moore, 1999; Barnett and Crispin, 2000). Although posterior uveitis could be independent from anterior uveitis, the limit between these entities is diffuse, and regularly all of them are affected together (Wilkie, 1994; Collins and Moore, 1999; Barnett and Crispin, 2000).

The posterior segment involvement may decrease the probability of maintaining a healthy ocular globe, and in this case, therapeutic agents should be systemically administered in order to reach the posterior segment. Although uveitis in cats may occur as a result of corneal ulceration, penetrating wound, or blunt trauma, infectious systemic diseases are the most frequent causes, due to the highly vascular nature and inmunosensibility of the uveal tract. According to previously published reports, approximately 30% to 62% of affected cats had no identifiable concurrent systemic disease (Gemensky et al., 1996; Chavkin et al., 1992). Feline immunodeficiency virus (FIV), feline infectious peritonitis (FIP), feline leukemia virus (FeLV), *Toxoplasma gondii*, *Cryptococcus neoformans*, and *Histoplasmacapsulatum* are the most frequent infectious agents implicated in the disease (Peiffer and Wilcock, 1991; Lappin et al., 1992; Glaze and Gelatt, 1999; Barnett and Crispin, 2000; Lappin, 2000). However, *Blastomycesdermatitidis*, *Coccidioidesimmitis* and *Candida albicans* (Lappin et al., 1992; Glaze and Gelatt, 1999; Barnett and Crispin,

2000; Lappin, 2000), *Bartonellahenselae* (Lappin et al., 1999), feline herpes virus 1 (FHV 1) (Lappin, 2000), *Cuterebra*spp (Johnson et al., 1988; Harris et al., 2000), *Mycobacterium bovis, Mycobacterium tuberculosis* and *Mycobacterium avium*(Gunn-Moore et al., 1996), and *Metastrongylidae*(Glaze and Gelatt, 1999) were also associated with feline uveitis (Table 1). However, the underlying cause is not identified in a substantial number of cats with uveitis (Lappin, 2000). While the advance of diagnostic techniques has allowed reducing the percentage of idiopathic uveitis, the majority of them are not differentiated and are classified as idiopathic (Glaze and Gelatt, 1999).

Table 1. Most frequent causes of feline uveitis

Exogenous	Endogenous
Trauma	Infectious
Corneal ulceration	Feline immunodeficiency virus (FIV)
Penetrating wounds	Feline leukemia virus (FeLV)
	Feline infectious peritonitis (FIP)
	Feline herpesvirus 1 (FHV 1)
	Cryptococcus neoformans
	Histoplasmacapsulatum
	Coccidioidesimmitis
	Blastomycesdermatitidis
	Candida albicans
	Fungidematiáceos
	Bartonellahenselae
	*Micobacteryum*spp.
	Parasitic
	Toxoplasma gondii
	Ophthalmomyiasis*Cuterebra*spp.
	Lens-induced uveitis (LIU)
	Idiopathic
	Neoplastic

In general terms, uveitis starts with a local lesion in the blood-ocular barrier (BOB) (Barnett and Crispin, 2000). The acute response to uveal damage is a brief vasoconstriction followed by a prolonged dilation of the arteriola-capilar-venule unit, which leads to an alteration of the vascular permeability. Although all the iridal vasculature has the same endothelial intercellular junctions (Glaze and Gelatt, 1999), feline iridal capillaries have discontinuous endothelial cells (Bellhorn, 1980) which favour the development of uveitis in this species. Therefore, proteins, erythrocytes (hyphema), and leukocytes (hypopyon) leave the vascular lumen and accumulate into the aqueous humor (AH), provoking anterior chamber turbidity (Tyndall effect). In addition, proteins and pigments from macrophages and pigmented stromal cells of the iris (Collins and Moore, 1999) can focally accumulate on the corneal endothelium (keratic precipitates), gravitate in the ventral anterior chamber (hypopyon) or infiltrate in the hypervascularised iris (rubeosisiridis), generating iridal swelling (Barnett and Crispin, 2000). Moreover, if the initial bleeding is extensive or continuous, a red blood cell complete filling of the anterior chamber can occur (Glaze and Gelatt, 1999). The intraocular inflammation could also provoke hyperemia in anterior ciliary and conjunctivae vessels (Glaze and Gelatt, 1999; Barnett and Crispin, 2000). Acute inflammation of the anterior uvea provokes pain, manifested by blepharospasm, photophobia, excessive tearing, and retraction of the eyeball (enophthalmos), with passive or active protrusion of the third eyelid (Barnett and Crispin, 2000).

The decrease in intraocular pressure (IOP) which is consequence of a AH production reduction provoked by cyclitis (Barnett and Crispin, 2000), or an increase in the uveoescleral flow mediated by prostaglandins (PGs) (Glaze and Gelatt, 1999), is an early indicator of uveitis (Glaze and Gelatt, 1999). In chronic processes, ciliary body fibrosis, atrophy, or both contribute to the decrease of the secretor function and eye hypotony (Glaze and Gelatt, 1999).

Pupillary constriction or miosis is induced by PGs and other inflammatory mediators, by direct action on the sphincter muscle. However, anterior or posterior synechiae kept contracted or moderately dilated pupils, which result in a fixed or deformed pupil (Glaze and Gelatt, 1999).

Corneal edemaobserved in some cases, is the result of corneal endothelial cell dysfunction induced by inflammatory mediators localized in the AH (Collins and Moore, 1999). These cells are involved in maintaining a low concentration of water in the cornea through a Na^{+}-K^{+} ATPase pump (Gum et al., 1999). In chronic cases, edema is caused by the leakage of fluid from deep perilimbal anterior ciliary vessels that invade the cornea towards the stroma (Barnett and Crispin, 2000).

Active chorioretinitis is characterized by cell infiltration and edema, observed in the ophthalmoscopic examination as grayish foci. The blood-retinal barrier (BRB) alteration produces edema and retinal detachment, perivascular cuffing, and bleeding. In the inactive chorioretinitis, ophthalmic signs are: alteration of tapetal coloration, hyperpigmentation or hypopigmentation of the retinal pigment epithelium (RPE) and tapetalhy-perreflectivity due to retinal thinning (Barnett and Crispin, 2000).

Complications associated to anterior or posterior uveitis are directly related with the duration and causes of the uveal inflammation (Glaze and Gelatt, 1999). Anterior synechiae occur if the iris is adhered to the corneal endothelium. If they are extensive, posterior synechiae may occlude the flow of AH from the posterior chamber through the pupil into the anterior chamber, causing iris bombe (anterior ballooning of the iris), and secondary glaucoma (Cassin and Rubin, 2001). Preiridalfibrovascular membranes may be formed on the anterior surface of the iris. Clinically, the term rubeosisiridis is applied, because the neovascular membrane on the anterior iridal surface causes a reddish cast to the iris (Cassin and Rubin, 2001).

Chronic inflammation often induces cataracts, presumably by the diffusion of inflammatory mediators across the lens capsule, causing lens epithelial metaplasia, necrosis, or posterior migration, as well as lens fiber degeneration, liquefaction, and necrosis (Eagle and Spencer, 1996; Streeten, 2000).

Although lens luxation seems to be relatively rare in dogs (Curtis et al., 1983), chronic uveitis is a frequent cause of lens luxation in cats because zonularfibers can be degraded by inflammatory products in the AH (Olivero et al., 1991). Secondary glaucoma may occur due to pupillary block as a result of iris bombe or lens luxation or may be attributable to occlusion of the iridocorneal angle by peripheral anterior synechiae (Stiles and Towsend, 2007). The final consequence of the inflammatory process could be buphtalmus or eyeball atrophy (*pthisisbulbis*) (Glaze and Gelatt, 1999).

Except in LIU and neoplasic uveitis, there are no specific ophthalmologic changes, particularly when the anterior uveal tract is affected (Lappin et al., 1992). However, some signs are frequently associated with specific causes (Glaze and Gelatt, 1999). Mutton fat keratic precipitates occur most frequently in FIP and FeLV uveitis, while those signs caused by FIP include haemorrhage (Glaze and Gelatt, 1999).

FIV and FeLV uveitis are associated with iridal nodules, similarly to white to pink masses in anterior chamber or attached to the anterior portion of the iris in uveitis associated with FeLV (Glaze and Gelatt, 1999). Infiltrated cells in anterior vitreous occur more frequently with FIV, as well as the formation of synechiae, cortical cataracts and glaucoma in chronic uveitis associated with this disease (Glaze and Gelatt, 1999). In general, FIP involves panuveitis or panophthalmitis with vitritis, chorioretinitis, marked anterior uveitis, severe and diffuse corneal edema and inflammatory retinal detachments (Glaze and Gelatt, 1999).

In addition, anterior uveitis provoked by FeLV can progress to a diffuse inflammation of the eyeball or a tumour formation (Glaze and Gelatt, 1999).

Despite that uveitis is one of the main causes of eye morbidity and loss of visual functions, the complexity of the biochemical and immune mechanisms involved in its generation and development remain incompletely elucidated. Several lines of evidence support that uveitis is due to damage generated by infiltrated leukocytes which release cytokines (de Vos et al., 1994; Planck et al., 1994) and other inflammatory chemical mediators, like arachidonic acid (AA) metabolites (Bellot et al., 1996), reactive oxygen species (ROS) (Rao, 1990), NO (Hashida et al., 2000), and superoxide anion (O_2^-) (Jacquemin et al., 1996), among many others. Oxygen-derived free radicals have been implicated in the inflammation associated with endotoxin-induced uveitis (EIU) (Satici et al., 2003).

Free radicals initiate retinal lipid peroxidation, and the resultant hydroperoxides serve as amplification factors, such as chemotaxis of inflamematory cells (Goto et al., 1991; Wu et al., 1992). Oxidative stress in photoreceptor mitochondria appears to be an early event in the development of uveitis, which initiates the subsequent irreversible retinal damage by attracting phagocytic cells to the initial site of injury (Saraswathy and Rao, 2008). In fact, there is evidence that free radical scavengers and antioxidants can act as anti-inflammatory agents, protecting the eye from inflammation-mediated tissue damage (Bosch-Morell et al., 2002).

Prostaglandins (PGs) can be synthesized through the activities of two cyclooxygenase (COX) isoforms. COX-1 is constitutively expressed in most tissues and its activity provides for the relative small amounts of PGs required for the mediation and modulation of normal physiological functions. In inflammatory conditions, COX-2 is rapidly induced by cytokines, growth factors, and LPS, and its enzymatic activity accounts for the large amounts of PGs produced during inflammation (Masferrer and Kulkarni, 1997). PGs in small doses administered topically or intraocularly reproduce some of the injury and inflammation responses, such as hyperaemia, miosis, breakdown of the BAB and ocular hypertension.

In addition, PGE_2 topically administered together with histamine (but not the individual components) cause cellular infiltration and produce edema in conjunctival tissues (Bhattacherjee, 1980). Moreover, it was shown that AA metabolites regulate vascular permeability, chemotaxis, and contribute to uveitis amplification (Bhattacherjee, 1980; Howes and McKay, 1976).

CURRENT THERAPY FOR FELINE UVEITIS

The early diagnosis and treatment implementation of uveitis are critical, due to the close anatomical relationship of the uvea with components directly implicated in visual functions. At present, the therapy for feline uveitis is essentially directed towards pain reduction and to avoid lesions of ocular tissues. The agents used to attain those goals include topical mydriatics, topical (and, in select cases, systemic) corticosteroids, and nonsteroidal anti-inflammatory drugs (NSAIDs).

The most often used mydriatic agent is atropine (1%) ointment or solution. This is a selective, reversible, direct-acting anticholinergic agent. The topical administration of atropine results in pharmacologic blockage of the sphincter muscles of the iris and ciliary body, leading to pupillary dilation and cycloplegia (Klauss and Constantinescu, 2004). The consequent pupillary dilation decreases the incidence of posterior synechiae. The cycloplegia greatly lessens the pain associated with ciliary spasm. The use of the ointment is indicated in felines, because the agent has a bitter taste, and cats may have particularly marked reactions (Gelatt et al., 1973).

Corticosteroids are frequently used to suppress inflammation because they reduce the production of metabolites of the AA cascade (Holmberg and Maggs, 2004). To control anterior uveitis, topical application is often the preferred route because it allows for high local drug concentrations, reducing systemic side effects (Holmberg and Maggs, 2004). Prednisolone acetate (1%) ophthalmic suspension and dexamethasone (0.1%) are able to penetrate an intact corneal epithelium (Wilkie, 1990). The use of glucocorticoids, such as prednisone, prednisolone, or dexamethasone to suppress inflammation within the posterior segment (chorioretinitis) requires a systemic administration or, occasionally, immunosuppressive doses (Holmberg and Maggs, 2004).

The NSAIDs can be particularly useful in cases of mild anterior uveitis, as adjunctive therapy when combined with topical corticosteroids, and to control posterior segment inflammation when an infectious aetiology cannot be completely discarded. NSAIDs block the conversion of AA to PGs (Giuliano, 2004), suppress polymorphonuclear cell locomotion and chemotaxis (Perianin et al., 1985), decrease expression of inflammatory cytokines (Leonardi et al., 2000), and may act as free radical scavengers (Flach, 1992). Only a few NSAIDs have been evaluated regarding their efficacy in controlling ocular inflammation. Flunixinmeglumine and aspirin have been shown to stabilize the BAB in experimental models of uveitis (Millichamp et al., 1991; Ward et al., 1992); flunixinmeglumine had good effects, whereas the effects of aspirin were moderate when compared with placebo. The relative efficacies have been studied in an anterior chamber paracentesis model of uveitis in dogs, and the order of BAB stabilizing efficacy was as follows: diclofenac>flurbiprofen> suprofen>tolmetin which was similar to placebo (Ward, 1996). However,theimmunosuppressive effect of corticoids may contribute to the development of the systemic disease and their chronic use could induce cortisonic glaucoma (Zhan et al., 1992). NSAIDs may delay the reparation process and coagulation, provoke acute renal insufficiency, gastrointestinal haemorrhage or ulceration (Wilkie, 1994; Barnett and Crispin, 2000), and have the risk of bone marrow suppression and haemorrhage in cats (Powell and Lappin, 2001). Moreover, it has been shown that only 33% of animals with uveitis associated with systemic diseases respond to this treatment (Gemensky et al., 1996).

There are several common features for uveitis among different species. However, many antiuveitic medications that are successfully used in humans and dogs are not very well tolerated by cats and, therefore, the selection of recommended drugs is limited in this species. Acquiring this knowledge relies heavily on the use of experimental models particularly developed in this species.

EXPERIMENTAL MODELS OF FELINE UVEITIS

An experimental model of feline uveitis would greatly facilitate the understanding of the cellular events involved in ocular inflammation as well as the development of new therapeutic strategies for uveitis in cats. In that context, techniques for experimental induction of uveitis in cats include inoculations of *T gondiivia*the carotid artery (Davidson, 1993). In that context, ocular toxoplasmosis in cats was experimentally induced *via* inoculation of a low number of organisms from a mildly virulent of *Toxoplasma gondii* into the common carotid artery. However, the pathogenicity of *Toxoplasma gondii* depends on the virulence of the strain, which could make difficult to reproduce the experimental model in different laboratories. Moreover, a mild inflammation in the anterior segment was observed following inoculation with *Toxoplasma gondii*in cats, whereas naturally occurring uveitis generally involves a more generalized ocular inflammation. Furthermore, experimentally induced ocular toxoplasmosis was associated with the breakdown of the retinal pigment epithelium cell barrier, whereas disruption of both anterior and posterior BOB should be modeled in order to more closely reproduce feline uveitis. Finally, ocular toxoplasmosis is a specific form of ocular disease, whereas an "ideal" experimental model of feline uveitis should reproduce pathological consequences caused by a greater variety of etiopathogenic agents.

EIU is an animal model of acute ocular inflammation induced by administration of LPS, a component of Gram-negative bacterial outer membranes (Rosembuan et al., 1980; Forrester et al., 1980; Csukas et al., 1990). LPS enhances the expression of various inflammatory mediators, such as interleukin (IL)-6 (Hoekzema et al., 1991),tumor necrosis factor (TNF)-α (Chi et al., 2004), PGE_2 (Rao et al., 1987) and monocyte chemotactic protein (MCP)-1 (de Vos et al., 1994), as well as the production of nitric oxide (NO) (Rao 1990), which contribute to the inflammation process, resulting in the breakdown of the BOB and leukocyte infiltration. Although EIU was originally used as a model of anterior uveitis, increasing evidences show that it also involves inflammation of the posterior segment of the eye with recruitment of leukocytes that adhere to the retinal vasculature and infiltrate the vitreous cavity (Kurihara et al., 2006; Yomura et al., 2007). In this vein, we have developed an experimental model of EIU in cats (Del Sole et al., 2008). For this purpose, we used a unilateral intravitreal injection of LPS because this maneuver avoids development of systemic inflammation and allows the use of the contralateral eye as a control, thereby decreasing the risk of treatment-related death, and minimizing the number of animals. The clinical consequences of LPS injection during a 45-day period were assessed by use of a scoring system and could be apportioned to approximately 3 phases: from days 1 through 12, from days 15 through 29, and from day 35 until the end of the study (day 45), respectively. In the first phase, mostly signs of changes in the anterior segment of the LPS-injected eyes were observed, whereas in the last phase, sequels in the posterior segment were most noticeable. Few studies of EIU in several species have previously addressed the temporal course of ocular alterations after LPS injection, and largely only short-term (from 24 hours to 7 days) consequences were reported. Our study indicates that a single intravitreal injection of LPS in cats induces ocular alterations compatible with feline uveitis, and that the duration of these effects is longer than previously estimated.

Regardless of the administration method used, the injection of LPS is known to provoke an increase in protein concentration and cell number in AH in rabbits, rats, and mice (Rosenbaum et al., 1980; Csukas et al., 1990; Kogiso et al., 1992). As expected, the intravitreal injection of LPS in cats provokes a significant increase in both parameters, supporting a disruption of the BOB integrity. In addition, significant reductions in ERG a- and b-wave amplitudes were detected in LPS-treated eyes, suggesting that sequels of ocular inflammation induced by LPS could be located in both outer and middle regions of the retina. In agreement with this suggestion, attenuation of the ERG a- and b-wave amplitudes in various animals including humans with uveitis has also been observed (Hamasaki et al., 1993; Kim et al., 1999; Horio et al., 2001). Moreover, at 45 days post-injection of LPS, retinal lesions formed by inward foldings of photoreceptor external segments and lymphocytes were observed in the inner region of the retina, indicating damage to the outer, middle, and inner regions of the cat retina. Moreover, similar to findings in cats with naturally occurring uveitis, IOP was significantly lower in LPS- than in vehicle-injected eyes. In summary, these results indicates that experimental uveitis induced *via* a single intravitreal injection of LPS in cats causes clinical, biochemical, histological, and functional changes which are consistent with the main features of naturally occurring feline uveitis (Del Sole et al., 2008). Thus, this model may be a useful tool in the investigation of new treatment strategies for uveitis in cats.

NO AND UVEITIS

Studies from animal models of uveitis and human patients with uveitis implicate reactive oxygen/nitrogen radicals in the inflammatory process. Parks et al. (1994) demonstrated that multiple intraperitoneal injections of N-omega-nitro-L-arginine methyl ester (L-NAME), a well-known NO synthase (NOS) inhibitor, reduced clinical signs of EIU, suggesting that NO could participate in the pathogenesis of experimental uveitis as a proinflammatory mediator. Furthermore, it has been shown that the hemodynamic and vascular permeability changes associated with EIU in the rat are in part due to increased production of NO. High levels of nitrite in the AH and vitreous of rats at the time of maximal ocular inflammation following LPS injection, and the presence of a calcium independent NOS activity in the anterior segment during EIU (Mandai et al., 1994) have been demonstrated. In addition, reverse transcriptase-PCR assays show an increased expression of the inducible isoform of NOS (iNOS) mRNA in the iris-ciliary body and in the retina of rats with experimental uveitis (Goureau et al., 1995).

In situ hybridization revealed that epithelial cells of the iris-ciliary body and cells infiltrating (macrophages, polymorphonuclear leukocytes) the anterior segment and the retina are the major source of NO (Jacquemin et al., 1996). Moreover, a significant correlation between nitrite levels in the AH and clinical signs of EIU was observed. The efficacy of treatment with NOS inhibitors in EIU in rat or rabbit suggests that treatments that reduce NO production or NO action would be beneficial and would provide new prospects for uveitis treatment (Becquet et al., 1997).

Ocular Functions of Melatonin

Systemic administration of melatonin has been found to produce significant changes in anterior and vitreous chamber depth, suggesting that melatonin may play a role in ocular growth and development (Rada and Wiechmann, 2006). Retinal melatonin acts as a neuromodulator that mediates dark adaptive regulation of retinomotor movements (Pierce and Besharse, 1985). The expression of melatonin type 1 receptors (MT1) in most dopaminergic amacrine cells in human retina implicates melatonin in the modulation of dopaminergic function (Scher et al., 2003). Both dopamine and melatonin are key signaling agents in the regulation of retinal rhythmicity. These two substances are mutually inhibitory, acting as signals for day and night, respectively. It was demonstrated that picomolar concentrations of melatonin selectively inhibit the calcium-dependent release of dopamine from rabbit retina (Dubocovich, 1983). The action of melatonin on the rod pathway at the level of horizontal and amacrine cells has been proposed as a unique mechanism by which the retina adapts to low light intensities (Scher et al., 2002). A correlation between melatonin levels and the electroretinographic response has been shown in human studies which suggest that daily melatonin and electroretinogram (ERG) cycling are associated (Rufiange et al., 2002). On the other hand, melatonin decreases retinal cAMP accumulation (Faillace et al., 1994) and regulates the activity of the retinal glutamate/glutamine cycle activity (Sáenz et al., 2004).

Melatonin As a RetinalAntioxidant

The possibility that melatonin could detoxify ROS was originally suggested by Ianas et al. (1991). Shortly thereafter, Poeggeler et al. (1993) and Reiter and coworkers (1994) using spin trapping and electron resonance spectroscopy, demonstrated melatonin's capacity to directly scavenge the highly reactive hydroxyl radical. Since then, several reports have shown that melatonin acts as a free radical scavenger and an efficient antioxidant (Hardeland et al., 1995; Reiter et al., 1997; 1998; 2000; Pandi-Perumal et al., 2006). Not only melatonin, but also several of its metabolites generated during its free radical scavenging action act as antioxidants (Tan et al., 2007). The kynurenic pathway of melatonin metabolism includes a series of radical scavengers with the possible sequence:

Melatonin → cyclic 3-hydroxymelatonin →N^1-acetyl-N^2-formyl-5-methoxykynuramine (AFMK) →N^1-acetyl-5-methoxykynuramine.

In the metabolic step from melatonin to AFMK, up to four free radicals can be consumed (Adler et al., 1997; Guenther et al., 2005; Hardeland, 2005; Tan et al., 2007). Because of this pathway, melatonin's efficacy as an antioxidant is greatly increased. Melatonin has been shown to protect against oxidative stress in various, highly divergent experimental systems. Melatonin's functions as an antioxidant include: a) direct free radical scavenging, b) stimulation of antioxidative enzymes, c) increasing the efficiency of mitochondrial oxidative phosphorylation and reducing electron leakage (thereby lowering free radical generation), and d) augmenting the efficiency of other antioxidants (Reiter et al., 2003).

Melatonin has been shown to scavenge free radicals generated in mitochondria, reduce electron leakage from the respiratory complexes and improve ATP synthesis (Acuña-Castroviejo et al., 2003; Leon et al., 2005). Moreover, melatonin preserves mitochondrial glutathione levels, thereby enhancing the antioxidant potential (Leon et al., 2004).

The retina is particularly susceptible to oxidative stress because of its high consumption of oxygen, its high proportion of polyunsaturated fatty acids, and its direct exposure to light. It was demonstrated that melatonin decreases lipid peroxidation of polyunsaturated fatty acids located in rod outer segment membranes (Guajardo et al., 2003), it may protect retinal pigment epithelial cells from oxidative stress (Liang et al., 2004), and it prevents retinal oxidative damage from ischemia/reperfusion injury (Celebi et al., 2002). In cultures of human retinal neuronal cells subjected to experimental hypoxanthine/xanthine oxidase -induced injury, melatonin is able to rescue retinal neurons (Lee et al., 2001). Furthermore, in dark-adapted single photoreceptors isolated from the frog retina after illumination with saturating light, melatonin at picomolar and low nanomolar concentrations was shown to be 100 times more potent in inhibiting light-induced oxidative processes than vitamin E (Marchiafava and Longoni, 1999). Recently, we have shown that melatonin increases superoxide dismutase activity and reduced glutathione levels, whereas it decreases lipid peroxidation in the rat retina (Belforte et al., 2010). Taken together, these results strongly support the role of melatonin as an effective retinal antioxidant.

Melatonin and the Retinal Nitridergic Pathway

NO is a highly reactive short-lived radical species which plays an important role as an inter and intracellular messenger in the central nervous system. NO is generated by a family of enzymes called NOS that oxidate L-arginine to L-citrulline producing NO. NO acts as an endogenous activator of a soluble guanylylcyclase, and thereby it increases cGMP levels.

There is evidence from different studies supporting that NO is involved in the regulation of several retinal processes. NO activates a retinal soluble guanylylcyclase, and modulates a Ca^{2+} channel and a voltage-independent conductance in the inner segment of photoreceptors (Kurenny et al., 1994). Effects of NO were also recorded on hemi-gap junction channels in horizontal cells of the catfish retina (DeVries and Schwartz, 1992), on cGMP-gated channels in retinal ganglion cells (Ahmad et al., 1994) and ON-bipolar cells (Nawy and Jahr, 1990). Intracellular application of L-arginine prevents the prolongation of the light responses in frog isolated rods (Tsuyama et al., 1993) and it blocks gap junctions between horizontal cells in the turtle retina (Miyachi et al., 1990).

In the retina as in other tissues, NO can be neuroprotective and neuro-destructive, depending on the redox state and on its concentration (Kashiwagi et al., 2001). NO can react with superoxide producing peroxynitrite, a powerful and deleterious oxidant which triggers apoptotic cell death (Sennlaub et al., 2002). Moreover, in a model of oxygen-induced retinopathy, it was demonstrated that iNOS is upregulated in response to tissue ischemia, is cytotoxic and inhibits vascular repair (Stevenson et al., 2010). In addition, a significant activation of the retinal nitridergic pathway was observed in hypertensive eyes (Belforte et al., 2007).

We have shown that in isolated conditions, melatonin may directly react with NO yielding at least one stable product, N-nitrosomelatonin, which was characterized by nuclear magnetic resonance (NMR) and X-ray spectroscopy (Turjanski et al., 2000a; 2000b). In addition to a direct scavenging of NO, low concentrations of melatonin significantly decrease hamster retinal NOS activity. Melatonin behaves as a non-competitive inhibitor of this enzyme, since it decreases the Vmax, with no changes on the Km (Sáenz et al., 2002). This result agrees with the previously described effect of melatonin on NOS from other neural structures such as hypothalamus (Bettahi et al., 1996), cerebellum (Pozo et al., 1997), striatum, and cerebral frontal cortex (León et al., 1998). However, while the inhibitory effect of melatonin in those tissues was evident up to 1 nM, in the retina a much higher sensitivity to the methoxyindole was evident, as it was effective even at 1 pM. In addition, while in other systems melatonin is active when added to crude homogenates, in the hamster retina, it is only effective when the whole tissues are preincubated in the presence of the methoxyindole. The high sensitivity of this melatonin inhibitory effect together with the fact that the tissue integrity seems to be necessary, suggest that in the retina, melatonin modulates NOS activity through a receptor-mediated mechanism, but not through its binding to calmodulin. The reasons of this discrepancy are not clear. It should be noted that the retina, unlike the cerebellum, the hypothalamus, or the striatum, is able to synthesize melatonin. Being a highly lipophilic molecule, melatonin has the capability of crossing all physiological barriers. However, these results suggest that the effect of melatonin on NOS activity could be restricted to those cell types that co-express melatonin receptors and NOS.

Besides NOS, another limiting step in NO synthesis is the availability of its precursor, L-arginine, which depends on the presence and activity of a specific uptake system. Physiological concentrations of melatonin significantly decrease retinal L-arginine influx (Sáenz et al., 2002). The lower Vmax of this mechanism observed in the presence of the methoxyindole suggests that melatonin decreases the number of active transporter molecules, without changing their affinity. Since melatonin decreases both L-arginine influx and its conversion to NO, a decrease in L-arginine-induced response is expectable in the presence of the methoxyindole. Indeed, the increase in cGMP accumulation induced by L-arginine is significantly reduced by melatonin (Sáenz et al., 2002). Despite the fact that NO donors, such as sodium nitroprusside releases NO in a NOS- independent way, we showed that melatonin also decreases its effect on cGMP accumulation, probably through its capacity to directly scavenge NO (Sáenz et al., 2002). In agreement with these results, it was shown that uncontrolled NO elevation causes morphological and nuclear changes in the retina and that melatonin signifycantly suppresses the NO induced increase in mean inner retinal layer thickness, nuclei in the inner nuclear layer, as well as the number of retinas expressing hyperchromatic expression, and apoptotic ganglion cells (Siu et al., 2004). In addition, it was demonstrated that melatonin reduces NO-induced lipid peroxidation in rat retinal homogenates (Siu et al., 1999). Peroxynitrite is a toxic oxidant formed from the reaction of superoxide and NO under conditions of inflammation and oxidant stress, and it was shown that melatonin is a scavenger of peroxynitrite (Gilad et al., 1997). Furthermore, it was demonstrated that melatonin is protective against oxidative damage in situations where NO is known to account for molecular destruction (Cuzzocrea et al., 2000).

Besides the already mentioned effects, several other mechanisms could be involved in the anti-inflammatory effect of melatonin, like inhibition of activity and expression of COX (Martínez-Campa et al., 2009), and PG release (Pawlikowski et al., 1984), inhibition of the rolling and adherence of leukocytes to endothelial layers, limiting cell migration (Markus et al., 2007), reduction of proinflammatory (Escames et al., 2006) and increase in anti-inflamatory cytokines (such as IL-1, IL-6 and IL-12), (Maestroni, 2001), decrease in vascular endothelial growth factor (Kaur et al., 2007) and chemokines levels (Park et al., 2007), and reduction of Fas ligand-mediated apoptosis, among many others, but up to now most of them were not confirmed at ocular level. Further investigations of the mechanisms by which melatonin affects its ocular target tissues, may unveil new concepts of ocular therapeutic.

Melatonin as a Therapeutic Resource for Uveitis Treatment

Kukner et al. (2006) and Sande et al. (2008) investigated the effectiveness of melatonin as a therapeutic strategy in uveitis. In the Kukner study, intravitreal injections of bovine serum albumin were used to induce uveitis in the guinea pig. Melatonin, vitamin E, or aprotinin were intraperitoneally administered 3 days after albumin injection. A significant increase in leptin expression occurs in the retina, choroid, sclera, and episclera; ganglion cells were edematous and inner plexiform layer thickness increased in uveitic eyes. In animals treated with melatonin, vitamin E, or aprotinin, leptin expression was similar to the control group and edema and histopathological changes were reduced (Kukner et al., 2006). Sande et al. studied EIU, another established experimental model for uveitis, giving a single intravitreal injection of LPS in the golden hamster. In this study, a pellet of melatonin was implanted subcutaneously 2 h before the intravitreal injection of LPS.

Twenty-four hours and 8 days after injection, the inflammatory response was evaluated in terms of: (i) integrity of BOB, (ii) clinical signs, (iii) histopathological studies, and (iv) retinal function. Melatonin reduced the leakage of proteins and cells in the anterior segment of LPS-injected eyes, decreased clinical signs, and protected the ultrastructure of BOBs. In animals injected with LPS, a remarkable disorganization of rod outer segment membranous disks was observed, whereas no morphological changes in photoreceptor outer segments were observed in animals treated with LPS plus melatonin. Furthermore, melatonin prevented the decrease in the ERG activity induced by LPS. Melatonin significantly abrogated the LPS-induced increase in retinal NOS activity, TNFα, and nuclear factor κB (NFκB) p50 and p65 subunits levels (Sande et al., 2008).

These results indicate that melatonin prevents clinical, biochemical, histological, ultrastructural, and functional consequences of experimental uveitis in hamsters, probably through an NFκB-dependent mechanism. Based on these results, we examined the effect of melatonin on experimetally-induced feline uveitis. For this purpose, melatonin was orally administered every 24 hr to a group of ten cats, from 24 h before until 45 days after intravitreal injections of vehicle or LPS (Del Sole et al., 2012).

Eyes were evaluated by means of clinical evaluation, intraocular pressure (IOP), BOB integrity (*via* measurement of protein concentration and cell content in samples of AH), ERG, and histological examination of the retinas. The treatment with melatonin significantly decreased clinical signs and prevented the reduction in IOP induced by LPS. In LPS-injected eyes, melatonin significantly preserved the BOB integrity, as shown by a decrease in the number of infiltrating cells and protein concentration in the AH. Mean amplitudes of scotopic ERG a- and b-waves were significantly reduced in eyes injected with LPS, whereas melatonin significantly prevented the effect of LPS. At 45 days after injection, LPS induced alterations in photoreceptors and at the middle portion of the retina, whereas melatonin preserved the retinal structure.

These results indicate that melatonin prevented clinical, biochemical, functional, and histological alterations induced by LPS injection in cats (Del Sole et al., 2012). Thus, melatonin might constitute a useful tool for the treatment of feline uveitis.

CONCLUSION

Uveitis, a disease with potentially blinding sequels, remains a challenging field to clinical and veterinarian ophthalmologists as the disease causes significant morbidity and the use of traditional forms of treatment is restricted by limited effectiveness and considerable side effects. Several lines of evidence support the involvement of oxidative damage and increased NO production as pathogenic mechanisms of uveitis. Melatonin by itself exhibits both antioxidant and antinitridergic properties, among many other anti-inflammatory effects. As mentioned before, ocular inflammation is mainly treated with topical and/or systemic application of corticosteroids. During long-term treatment with corticosteroids, however, care must be taken to guard against both ocular and systemic complications such as cataract, cortisonic glaucoma, diabetes, immunosuppressive effect, etc. Therefore, the establishment of additive anti-inflammatory approaches is desirable to decrease the rate and degree of these complications. The results obtained in hamsters and cats strongly support that melatonin, which lacks of adverse collateral effects even at high doses, could be a promissory resource in the management of uveitis. Alone or combined with corticosteroid therapy, the anti-inflammatory effects melatonin may benefit patients with chronic uveitis and decrease the rate and degree of corticosteroid-induced complications. Although in the mentioned studies (Del Sole et al., 2012) only the preventive effect of melatonin in feline experimental uveitis was demonstrated, this "relative weakness" does not preclude its clinical relevance. Indeed, uveitis is a common consequence of ocular surgery (cataract surgery and vitrectomy, among others). Thus, even as preventive strategy, the present results suggest that melatonin, a very safe compound for human and cat use, should be included in the ophthalmic therapeutic resources. In summary, melatonin appears as a potentially useful anti-inflammatory drug, particularly in ophthalmology, and as an alternative or eventually as complement to glucocorticoids, since it resulted to be highly effective and it does not present the side effects of these steroids.

REFERENCES

Acuña-Castroviejo D, Escames G, Leon J, et al. Mitochondrial regulation by melatonin and its metabolites. *Adv. Exp. Med. Biol.* 2003; 527:549-557.

Adler LJ, Gyulai FE, Diehl DJ, et al. Regional brain activity changes associated with fentanyl analgesia elucidated by positron emission tomography. *Anesth. Analg.* 1997; 84: 120-126.

Ahmad I, Leinders-Zufall T, Kocsis JD, et al. Retinal ganglion cells express a cGMP-gated cation conductance activatable by nitric oxide donors. *Neuron.*1994; 1: 155-165.

Barnett KC, Crispin SM. Tractouveal. In: Barnett KC, Crispin SM, eds. *OftalmologíaFelina atlas and texto.* Buenos Aires: Intermédica, 2000; 118-139.

Becquet F, Courtois Y, Goureau O. Nitric oxide in the eye: multifaceted roles and diverse outcomes. *SurvOphthalmol.* 1997; 42: 71-82.

Belforte N, Moreno MC, Cymeryng C, et al. Effect of ocular hypertension on retinal |nitridergic pathway activity. *Invest Ophthalmol Vis Sci.* 2007; 48: 2127-2133.

Belforte NA, Moreno MC, de Zavalía N, et al. Melatonin: a novel neuroprotectant for the treatment of glaucoma.*J. Pineal Res.*2010; 48: 353-364.

Bellhorn RW. Permeability of blood-ocular barriers of neonatal and adult cat to sodium fluorescein. *Invest Ophthalmol. Vis. Sci.* 1980; 19: 870-877.

Bellot JL, Palmero M, Garcia-Cabanes C, et al. Additive effect of nitric oxide and prostaglandin-E2 synthesis inhibitors in endotoxin-induced uveitis in the rabbit. *InflammRes.* 1996; 45: 203-208.

Bettahi I, Pozo D, Osuna C, et al. Melatonin reduces nitric oxide synthase activity in rat hypothalamus. *J. Pineal Res.* 1996; 20: 205-210.

Bhattacherjee P. Prostaglandins and inflammatory reactions in the eye. *Methods Find Exp. Clin. Pharmacol.*1980; 2: 17-31.

Bosch-Morell F, Roma J, Marin N, et al. Role of oxygen and nitrogen species in experimental uveitis: anti-inflammatory activity of the synthetic antioxidant ebselen. *Free RadicBiol Med.*2002; 33: 669-675.

Cassin B, Rubin M. *Dictionary of eye terminology.*4th edition. Gainesville (FL): Triad Publishing Company; 2001.

Celebi S, Dilsiz N, Yilmaz T, Kukner AS. Effects of melatonin, vitamin E and octreotide on lipid peroxidation during ischemia-reperfusion in the guinea pig retina. *Eur. J. Ophthalmol.*2002; 12: 77-83.

Chavkin M, Lappin M, Powell C, et al. Seroepidemiologic and clinical observations of 93 cases of uveitis in cats. *Prog. Vet. Comp. Ophthalmol.*1992; 2: 29-36.

Chi ZL, Hayasaka S, Zhang XY, et al. Effects of rolipram, a selective inhibitor of type 4 phosphodiesterase, on lipopolysaccharide-induced uveitis in rats. *Invest Ophthalmol. Vis. Sci.* 2004; 45: 2497-2502.

Collins B.K.; Moore C.P. (1999). Disease and surgery of the canine anterior uvea, pp151-81. *En*: K.N. Gelatt (ed.) Veterinary Ophthalmology, TerceraEdición. Lippincott Williams Wilkins. Philadelphia.

Csukas S, Paterson CA, Brown K, et al. Time course of rabbit ocular inflammatory response and mediator release after intravitreal endotoxin. *Invest Ophthalmol Vis. Sci.* 1990; 31: 382-387.

Curtis R, Barnett K, Lewis S. Clinical and pathological observations concerning the aetiology of primary lens luxation in the dog. *Vet. Rec.* 1983; 112: 238-246.

Cuzzocrea S, Costantino G, Gitto E, et al. Protective effects of melatonin in ischemic brain injury.*J. Pineal Res.* 2000; 29: 217-227.

Davidson MG, Lappin MR, English RV, et al. A feline model of ocular toxoplasmosis. *Invest Ophthalmol Vis. Sci.* 1993; 34: 3653-3660.

deVos AF, Klaren VN, Kijlstra A. Expression of multiple cytokines and IL-1RA in the uvea and retina during endotoxin-induced uveitis in the rat. *InvestOphthalmol Vis. Sci.* 1994; 35: 3873-3883.

Del Sole MJ, Sande PH, Felipe AE, Fernandez DC et al. Characterization of uveitis induced by use of a single intravitreal injection of bacterial lipopolysaccharide in cats. *Am. J. Vet. Res.* 2008; 11: 1487-1495.

Del Sole MJ, Sande JM, Fernandez DC, et al. Therapeutic benefit of melatonin in experimental feline uveitis.*J. Pineal Res.* 2012; 52: 29-37.

DeVries SH, Schwartz EA. Hemi-gap junction channels in solitary cells of the catfish retina. *J. Physiol.* 1992; 445: 201-230.

Dubocovich ML. Melatonin is a potent modulator of dopamine release in the retina. *Nature.*1983; 306: 782-784.

Eagle R, SpencerW. Lens. In: SpencerW, editor. *Ophthalmic pathology*. 4th edition. Philadelphia: WB Saunders; 1996. p. 372-427.

Escames G, Acuña-Castroviejo D, López LC, et al. Pharmacological utility of melatonin in the treatment of septic shock: experimental and clinical evidence. *J. Pharm. Pharmacol.*2006; 58: 1153-1165.

Faillace MP, Sarmiento MI, Siri LN, Rosenstein RE. Diurnal variations in cyclic AMP and melatonin content of golden hamster retina. *J. Neurochem.* 1994; 62: 1995-2000.

Flach A. Cyclo-oxygenase inhibitors in ophthalmology. *Surv. Ophthalmol.*1992; 36: 259-284.

Forrester JV, Worgul BV, Merriam GR Jr. Endotoxin-induced uveitis in the rat. *Albrecht Von Graefes Arch. Klin. Ophthalmol.*1980; 213: 221-233.

Gelatt K, Boggess T, Cure T. Evaluation of mydriatics in the cat. *J. Am. Anim. Hosp. Assoc.* 1973; 9: 283-287.

Gemensky A, Lorimer D, Blanchard G. Feline uveitis: a retrospective study of 45 cases. *Presented at Annual Meeting of the American College of Veterinary Ophthalmologists* 1996.

Gilad E, Cuzzocrea S, Zingarelli B, Salzman AL, Szabó C. Melatonin is a scavenger of peroxynitrite. *Life Sci.* 1997; 60: 169-174.

Giuliano E. Nonsteroidal anti-inflammatory drugs in veterinary ophthalmology. *Vet. Clin. North Am. Small Anim. Pract.*2004; 34: 707-723.

Glaze MB, Gelatt KN. Feline ophthalmology. In: Gelatt KN, ed. *Veterinary ophthalmology.* 3th ed. Philadelphia: Lippincott Williams and Wilkins, 1998; 1018-1025.

Goto H, Wu GS, Gritz DC, et al. Chemotactic activity of the peroxidized retinal membrane lipids in experimental autoimmune uveitis. *Curr. Eye Res.* 1991; 10: 1009-1014.

Goureau O, Bellot J, Thillaye B, et al. Increased nitric oxide production in endotoxin-induced uveitis: reduction of uveitis by an inhibitor of nitric oxide synthase. *J. Immunol.* 1995; 154: 6518-6523.

Guajardo MH, Terrasa AM, Catala A. Protective effect of indoleamines on in vitro ascorbate-Fe2+ dependent lipid peroxidation of rod outer segment membranes of bovine retina. *J. Pineal Res.* 2003; 35: 276-282.

Guenther AL, Schmidt SI, Laatsch H, et al. Reactions of the melatonin metabolite AMK (N1-acetyl-5-methoxykynuramine) with reactive nitrogen species: formation of novel compounds, 3-acetamidomethyl-6-methoxycinnolinone and 3-nitro-AMK. *J. Pineal Res.* 2005; 39: 251-260.

Gum GG, Gelatt KN, Esson DW Physiology of the Eye. In: Gelatt KN, ed. *Veterinary ophthalmology.* 4th ed. Blackwell Publishing, Iowa, 2007; 149-182.

Gunn-Moore DA, Jenkins PA, Lucke VM. Feline tuberculosis: a literature review and discussion of 19 cases caused by an unusual mycobacterial variant. *Vet. Res.* 1996; 138: 53-58.

Hamasaki DI, Sato H, Santhanakrishnan S, Shinohara T. Correlation between the physiologic and morphologic changes in experimental autoimmune uveitis induced by peptide G of S antigen. *Invest Ophthalmol Vis. Sci.* 1993; 34: 1861-1871.

Hardeland R, Balzer I, Poeggeler B, et al. On the primary functions of melatonin in evolution: mediation of photoperiodic signals in a unicell, photooxidation, and scavenging of free radicals. *J. Pineal Res.* 1995; 18: 104-111.

Hardeland R. Antioxidative protection by melatonin: multiplicity of mechanisms from radical detoxification to radical avoidance. *Endocrine.*2005; 27: 119-130.

Harris BP, Miller PE, Bloss JR, Pellittieri PJ. Ophthalmomyasisinterna anterior associated with *Cuterebra*spp in a cat. *J. Am. Vet. Med. Assoc.* 2000; 216, 352-355.

Hashida M, Fukushima A, Zhang J, et al. Involvement of superoxide generated by polymorphonuclear leukocytes in endotoxin-induced uveitis. *GraefesArch. Clin. Exp. Ophthalmol.* 2000; 238: 359-365.

Hoekzema R, Murray PI, van Haren MA, et al. Analysis of interleukin-6 in endotoxin-induced uveitis. *Invest Ophthalmol. Vis. Sci.* 1991; 32: 88-95.

Holmberg B, Maggs D. The use of corticosteroids to treat ocular inflammation. *Vet. Clin. North Am. Small Anim. Pract.* 2004; 34: 693-705.

Horio N, Terasaki H, Yamamoto E, Miyake Y. Electroretinogram in the diagnosis of endophthalmitis after intraocular lens implantation. *Am. J. Ophthalmol.* 2001; 132: 258-259.

Howes EL, Jr, McKay DG. The effects of aspirin and indomethacin on the ocular response to circulating bacterial endotoxin in the rabbit. *Invest. Ophthalmol.* 1976; 15: 648-651.

Ianas O, Olinescu R, Badescu I. Melatonin involvement in oxidative processes. *Endocrinologie.* 1991; 29: 147-153.

Jacquemin E, De Kozak Y, Thillaye B, et al. Expression of inducible nitric oxide synthase in the eye from endotoxin-induced uveitis rats. *Invest Ophthalmol Vis. Sci.* 1996; 37: 1187-1196.

Johnson BW, Helper LC, Szajerski ME. Intraocular *Cuterebra* in a cat. *J Am Vet. Med. Assoc.* 1988; 193: 829-830.

Kashiwagi K, Iizuka Y, Tanaka Y, et al. Dual action of nitric oxide on purely isolated retinal ganglion cells. *Curr. Eye Res.* 2001; 23:233-239.

Kaur C, Sivakumar V, Yong Z, et al. Blood-retinal barrier disruption and ultrastructural changes in the hypoxic retina in adult rats: the beneficial effect of melatonin administration. *J. Pathol.* 2007; 212: 429-439.

Kim IT, Park SK, Lim JH. Inflammatory response in experimental Staphylococcus and *Pseudomonas*endophthalmitis. *Ophthalmologica* 1999; 213: 305-310.

Klauss G, Constantinescu G. Nonhypotensive autonomic agents in veterinary ophthalmology. *Vet. Clin. North Am. Small Anim. Pract.* 2004; 34: 777-800.

Kogiso M, Tanouchi Y, Mimura Y, et al. Endotoxin-induced uveitis in mice. 1. Induction of uveitis and role of T lymphocytes. *Jpn. J. Ophthalmol.* 1992; 36: 281-290.

Kükner A, Colakoğlu N, Serin D, et al. Effects of intraperitoneal vitamin E, melatonin and aprotinin on leptin expression in the guinea pig eye during experimental uveitis. *ActaOphthalmol. Scand.* 2006; 84: 54-61.

Kurenny DE, Moroz LL, Turner RW, et al. Modulation of ion channels in rod photoreceptors by nitric oxide. *Neuron.* 1994; 13: 315-324.

Kurihara T, Ozawa Y, Shinoda K, et al. Neuroprotective effects of angiotensin II type 1 receptor (AT1R) blocker, telmisartan, via modulating AT1R and AT2R signaling in retinal inflammation. *Invest Ophthalmol Vis. Sci.* 2006; 47: 5545-5552.

Lappin MR, Marks A, Greene CE, et al. Serologic prevalence of selected infectious diseases in cats with uveitis. *J. Am. Vet. Med. Assoc.* 1992; 201: 1005-1009.

Lappin MR. Feline infectious uveitis. *J. Feline Med. Surg.* 2000; 2: 159-163.

Lee MC, Chung YT, Lee JH, et al. Antioxidant effect of melatonin in human retinal neuron cultures. *ExpNeurol.* 2001; 172: 407-415.

León J, Acuña-Castroviejo D, Escames G, et al. Melatonin mitigates mitocondrial malfunction. *J. Pineal Res.* 2005; 38: 1-9.

León J, Acuña-Castroviejo D, Sainz RM, et al. Melatonin and mitochondrialfunction. *Life Sci.* 2004; 75: 765-790.

Leonardi A, Busato F, Fregona I, et al. Anti-inflammatory and antiallergic effects of ketorolac tromethamine in the conjunctival provocation model. *Br. J. Ophthalmol.* 2000; 84: 1228-1232.

Liang FQ, Green L, Wang C, et al. Melatonin protects human retinal pigment epithelial (RPE) cells against oxidative stress. *Exp. Eye Res.* 2004; 78: 1069-1075.

Maestroni GJ. The immunotherapeutic potential of melatonin. *Expert Opin. Investig Drugs.* 2001; 10: 467-476.

Mandai M, Yoshimura N, Yoshida M, et al. The role of nitric oxide synthase in endotoxin-induced uveitis: effect of N^G-nitro L-arginine. *Invest Ophthalmol. Vis. Sci.* 1994; 35: 3673-3681.

Marchiafava PL, Longoni B. Melatonin as an antioxidant in retinal photoreceptors. *J. Pineal Res.* 1999; 26: 184-189.

Markus RP, Ferreira ZS, Fernandes PA, Cecon E. The immune-pineal axis: a shuttle between endocrine and paracrine melatonin sources. *Neuroimmunomodulation.* 2007; 14: 126-133.

Martínez-Campa C, González A, Mediavilla MD, et al. Melatonin inhibits aromatase promoter expression by regulating cyclooxygenases expression and activity in breast cancer cells. *Br. J. Cancer.* 2009; 101: 1613-1619.

Masferrer JL, Kulkarni PS. Cyclooxygenase-2 inhibitors: a new approach to the therapy of ocular inflammation. *Surv. Ophthalmol.* 1997; 41: 35-40.

Millichamp N, Dziezyc J, Rhohde B, et al. Acute effects of anti-inflammatory drugs on neodymium:nyttriumaluminum garnet laser-induced uveitis in dogs. *Am. J. Vet. Res.* 1991; 52: 1279-1284.

Miyachi E, Murakami M, Nakaki T. Arginine blocks gap junctions between retinal horizontal cells. *Neuroreport.* 1990; 1: 107-110.

Nawy S, Jahr CE. Suppression by glutamate of cGMP-activated conductance in retinal bipolar cells. *Nature.* 1990; 346: 269-271.

Olivero DK, Riis RD, Dutton AG, et al. Feline lens displacement: a retrospective analysis of 345 cases. *Prog. Vet. Comp. Ophthalmol* 1991; 1: 239-244.

Pandi-Perumal SR, Srinivasan V, Maestroni GJ, et al. Melatonin: Nature's most versatile biological signal? *FEBS J.* 2006; 273: 2813-2838.

Park HJ, Kim HJ, Ra J, et al. Melatonin inhibits lipopolysaccharide-induced chemokine subfamily gene expression in human peripheral blood mononuclear cells in a microarray analysis. *J. Pineal Res.* 2007; 43: 121-129.

Parks DJ, Cheung MK, Chan CC, Roberge FG. The role of nitric oxide in uveitis. *Arch. Ophthalmol.* 1994; 112: 544-546.

Pawlikowski M, Juszczak M, Karasek E, et al. Melatonin inhibits prostaglandin E release from the medial basal hypothalamus of pinealectomized rats. *J. Pineal Res.* 1984; 1: 317-321.

Peiffer, R.L.; Wilcock, B.P. (1991). Histopathologic study of uveitis in cats: 139 cases (1978-1988). *J. Am. Vet. Med. Assoc.* 198, 135-138.

Perianin A, Roch-Arveiller M, Giround J, et al. In vivo effects of indomethacin and flurbiprofen on the locomotion of neutrophils elicited by immune and non-immune inflammation in the rat. *Eur. J. Pharmacol.* 1985; 106: 327-333.

Pierce ME, Besharse JC. Circadian regulation of retinomotor movements. I. Interaction of melatonin and dopamine in the control of cone length. *J. Gen. Physiol.* 1985; 86: 671-689.

Planck SR, Huang XN, Robertson JE, Rosenbaum JT. Cytokine mRNA levels in rat ocular tissues after systemic endotoxin treatment. *Invest Ophthalmol Vis Sci.*1994; 35: 924-930.

Poeggeler B, Reiter RJ, Tan DX, et al. Melatonin, hydroxyl radical-mediated oxidative damage, and aging: a hypothesis. *J. Pineal Res.* 1993; 14: 151-168.

Powell CC, Lappin MR. Diagnosis and treatment of feline uveitis. *CompCont Educ. Pract.* 2001; 23: 258-266.

Pozo D, Reiter RJ, Calvo JC, et al. Inhibition of cerebellar nitric oxide synhtase and cyclic GMP production by melatonin via complex formation with calmodulin. *J. Cell Biochem.* 1997; 65: 430-442.

Rada JA, Wiechmann AF. Melatonin receptors in chick ocular tissues: implications for a role of melatonin in ocular growth regulation. *Invest Ophthalmol. Vis. Sci.* 2006; 47: 25-33.

Rao NA, Patchett R, Fernandez MA, et al. Treatment of experimental granulomatous uveitis by lipoxygenase and cyclo-oxygenase inhibitors. *Arch. Ophthalmol.* 1987; 105: 413-415.

Rao NA. Role of oxygen free radicals in retinal damage associated with experimental uveitis. *Trans Am. Ophthalmol. Soc.* 1990; 88: 797-850.

Reiter RJ, Garcia JJ, Pie J. Oxidative toxicity in models of neurodegeneration: responses to melatonin. *Restor. Neurol. Neurosci.* 1998; 12: 135-142.

Reiter RJ, Guerrero JM, Escames G, et al. Prophylactic actions of melatonin in oxidative neurotoxicity. *Ann NY AcadSci.* 1997; 825: 70-78.

Reiter RJ, Tan DX, Mayo JC, et al. Melatonin as an antioxidant: biochemical mechanisms and pathophysiological implications in humans. *ActaBiochim. Pol.* 2003; 50: 1129-1146.

Reiter RJ, Tan DX, Osuna C, Gitto E. Actions of melatonin in the reduction of oxidative stress. A review. *J. Biomed. Sci.* 2000; 7: 444-458.

Reiter RJ, Tan DX, Poeggeler B, et al. Melatonin as a free radical scavenger: implications for aging and age-related diseases. *Ann. NY Acad. Sci.* 1994; 719: 1-12.

Rosenbaum JT, McDevitt HO, Guss RB, et al. Endotoxin-induced uveitis in rats as a model for human disease. *Nature.* 1980; 286: 611-613.

Rufiange M, Dumont M, Lachapelle P. Correlating retinal function with melatonin secretion in subjects with an early or late circadian phase. *Invest Ophthalmol Vis. Sci.* 2002; 43: 2491-2499.

Sáenz DA, Goldin AP, Minces L, et al. Effect of melatonin on the retinal glutamate/glutamine cycle in the golden hamster retina. *FASEB J.* 2004; 18: 1912-1913.

Sáenz DA, Turjanski AG, Sacca GB, et al. Physiological concentrations of melatonin inhibit the nitridergic pathway in the Syrian hamster retina. *J. Pineal. Res.* 2002; 33: 31-36.

Sande PH, Fernandez DC, Aldana Marcos HJ, et al. Therapeuticeffect of melatonin in experimentaluveitis. *Am. J. Pathol.* 2008; 173: 1702-1713.

Saraswathy S, Rao NA. Photoreceptor mitochondrial oxidative stress in experimentalautoimmuneuveitis. *Ophthalmic. Res.* 2008; 40: 160-164.

Satici A, Guzey M, Gurler B, et al. Malondialdehyde and antioxidant enzyme levels in the aqueous humor of rabbits in endotoxin-induced uveitis. *Eur. J. Ophthalmol.*2003; 13: 779-783.

Scher J, Wankiewicz E, Brown GM, Fujieda H. AII amacrine cells express the MT1 melatonin receptor in human and macaque retina. *Exp. Eye Res.* 2003; 77: 375-382.

Scher J, Wankiewicz E, Brown GM, Fujieda H. MT(1) melatonin receptor in the human retina: expression and localization. *Invest. Ophthalmol Vis. Sci.* 2002; 43: 889-897.

Sennlaub F, Courtois Y, Goureau O. Inducible nitric oxide synthase mediates retinal apoptosis in ischemic proliferative retinopathy. *J. Neurosci.* 2002; 22:3987-3993.

Siu AW, Ortiz GG, Benitez-King G, et al. Effects of melatonin on the nitric oxide treated retina. *Br J Ophthalmol.* 2004; 88: 1078-1081.

Siu AW, Reiter RJ, To CH. Pineal indoleamines and vitamin E reduce nitric oxide-induced lipid peroxidation in rat retinal homogenates *J Pineal Res.* 1999; 2: 122-128.

Stiles J, Townsend WM. Feline ophthalmology. In: Gelatt KN, editor. 4th edition, *Veterinary ophthalmology*, vol. 2. Ames (IA): Blackwell Publishing; 2007. p. 1117-1124.

Stevenson L, Matesanz N, Colhoun L, et al. Reduced nitro-oxidative stress and neural cell death suggests a protective role for microglial cells in TNFalpha-/- mice in ischemic retinopathy. *Invest. Ophthalmol Vis Sci.* 2010; 51: 3291-3299.

Streeten B. Pathology of the lens. In: Albert D, Jakobiec F, editors. *Principles and practice of ophthalmology*. Philadelphia: WB Saunders; 2000. p. 3685-749.

Tan DX, Manchester LC, Terron MP, et al. One molecule, many derivatives: a never-ending interaction of melatonin with reactive oxygen and nitrogen species? *J. Pineal Res.* 2007; 42: 28-42.

Tsuyama Y, Noll GN, Schmidt KF. L-arginine and nicotinamide adenine dinucleotide phosphate alter dark voltage and accelerate light response recovery in isolated retinal rods of the frog (Ranatemporaria). *Neurosci. Lett.* 1993; 149: 95-98.

Turjanski AG, Chaia Z, Doctorovich F, et al. N-Nitrosomelatonin. *ActaCrystallogr. C.* 2000; 56: 682-683.

Turjanski AG, Leonik F, Estrin DA, et al. Scavenging of NO by Melatonin. *J. Am. Chem Soc.* 2000; 122: 10468-10469.

Ward D, Ferguson D, Ward S, et al. Comparison of the blood-aqueous barrier stabilizing effects of steroidal and nonsteroidal anti-inflammatory agents in the dog. *Prog. Vet Comp. Ophthalmol* 1992; 2: 117-124.

Ward D. Comparative efficacy of topically applied flurbiprofen, diclofenac, tolmetin, and suprofen for the treatment of experimentally induced blood-aqueous barrier disruption in dogs. *Am. J. Vet. Res.* 1996; 57: 875-878.

Wilkie DA. Control of ocular inflammation. *Vet. Clin. North Am. Small AnimPract.* 1990; 20: 693-713.

Wilkie DA. Trastornos de la úvea. In: Kirk RW, Bonagura JD, eds. Terapéutica veterinaria de pequeños animales XI. Madrid: McGraw-Hill Interamericana, 1994; 1233-1240.

Wu GS, Sevanian A, Rao NA. Detection of retinal lipid hydroperoxides in experimental uveitis. *Free Radic. Biol. Med.* 1992; 12: 1927.

Yomura Y, Shoji Y, Asai D, et al. Direct, real-time, simultaneous monitoring of intravitreal nitric oxide and oxygen in endotoxininduced uveitis in rabbits. *Life Sci.* 2007; 80: 1449-1457.

Zhan GL, Miranda OC, Bito LZ. Steroid glaucoma: corticosteroid-induced ocular hypertension in cats. *Exp. Eye Res.* 1992; 54: 211-218.

INDEX

B

C

D

E

F

G

H

I

T